RESEARCH AND INNOVATION

a record of
THE WOLFSON TECHNOLOGICAL
PROJECTS SCHEME 1968–1981

Dr L. Rotherham, CBE, FRS
Former Vice-Chancellor, Bath University

with a Foreword and Postscript by
Lord Zuckerman, OM, KCB, FRS

CLARENDON PRESS · OXFORD
1984

Oxford University Press, Walton Street, Oxford OX2 6DP

London New York Toronto
Delhi Bombay Calcutta Madras Karachi
Kuala Lumpur Singapore Hong Kong Tokyo
Nairobi Dar es Salaam Cape Town
Melbourne Auckland

and associated companies in
Beirut Berlin Ibadan Mexico City Nicosia

Oxford is a trade mark of Oxford University Press

Published in the United States
by Oxford University Press, New York

British Library Cataloguing in Publication

Rotherham, L.
Research and innovation.—(Monographs on
science, technology and society)
1. Wolfson Technological Projects Scheme—
History
I. Title   II. Series
606'.041   T173.8
ISBN 0-19-858335-4

Set by DMB (Typesetting), Oxford
Printed in Great Britain at
The University Press Oxford
by David Stanford
Printer to the University

FOREWORD

by Lord Zuckerman, OM, KCB, FRS
Wolfson Trustee

This small book tells the story of the Wolfson Technological Projects Scheme from its inception in 1967 up to the start of the 1980s. The Scheme needs to be viewed in the context of the environment within which it emerged, and its future in the light of the changes that have occurred since then.

To many who were concerned with our manufacturing industries, the sixties were a time of disillusion, bewilderment, and even despair. Our national reputation for science was good, our universities had been expanded to increase the output of trained scientific manpower, the Government was spending immense sums on programmes of research and development, and government agencies were all hard at work. Yet our industries were steadily losing ground in world markets to foreign competitors, in particular, at the time, to the United States.

Many factors had contributed to this discouraging situation, and there were many views about how it could be corrected. Whatever else, a majority of those who had given their minds to the problem believed that if industry were to benefit from the enormous range of scientific talent available in our universities, the gap between universities and industry had to be bridged, and bridged quickly. No scheme to achieve this end was then in existence.

By 1967 the Wolfson Foundation had already devoted many millions of pounds to the support of higher education. It was therefore logical for the Foundation to embark on a policy of direct action in order to narrow the gap and to engage universities and industry in joint development projects.

When the Scheme started, we were to a considerable extent stepping into the dark, and it was my view that the Foundation would be lucky if one in twenty of the industrial-development projects which we supported in universities were to 'succeed'—a term which I and my fellow-Trustees were ready to interpret very broadly. Over the past fifteen years we have in fact done much better than that. Moreover, it has become clear that 'success' does not just mean developing a new product or a new process, or instituting an advisory service for local manufacturers. The forging of links between universities and industry benefits both sides, and not least in its effects on university teaching, and hence on the attitudes and ambitions of young graduates. It is in the help it has given to the achievement of this rapprochement that we must seek the major success of the Scheme. In all our universities it has helped to impart to young people a 'feel' for industry and commerce, and an interest in the innovative process and the generation of wealth. Today, with youth unemployment a major feature of the economic scene, this has become an issue of greater importance than we envisaged at the start.

The Foundation has administered the Scheme with the minimum of formality and with the flexibility needed to meet local conditions and changing situations. More recently, we have also instituted a programme of grants to help support trained scientists and engineers who were previously engaged on research and development but who have lost their jobs because of retrenchments in the companies by which they were employed. We are also taking a direct interest in studies whose purpose it is to present a picture of the pattern of future employment for Britian in an industrial world which is so rapidly being transformed by new technology.

The Trustees are indebted to Dr. Rotherham for the unstinting help that he has given in the administration of the Technological Projects Scheme. Not only has he helped in assessing the merits of the different proposals that were submitted to us; he also made it his business to follow the progress of those that were supported. His experience of manufacturing industry, of work in a government establishment, as a director of a national corporation, and as a Vice-Chancellor has proved invaluable.

We hope that this book, which takes stock of what has been achieved so far, and of the problems that have been encountered, will be of value to those who are now involved in corresponding efforts to relate university research and teaching to the economic regeneration of our country.

PREFACE

At the invitation of the Trustees of the Wolfson Foundation, I have set out in this book a record of a unique scheme, whose purpose is to help bridge the gap between university research workers on the one hand, and those responsible for innovation in industry on the other, and to which the Foundation has devoted some £17 million over the past sixteen years. The existence of the gap has been widely recognized, and for well over a hundred years has been a cause for concern. As is explained in the opening chapter, most of the measures to narrow it that have been introduced by government have been of a general kind—the allocation of funds to encourage basic research in the sciences and to increase the output of trained scientific manpower. The Wolfson Foundation has tried to tackle the problem in a more direct way.

There have, of course, been some universities in which scientists and engineers were never discouraged from entering into consultancies with manufacturing industry. But in general the practice used to be frowned upon. For example, it was common knowledge that up to the outbreak of the second world war in 1939, only one science professor in Oxford supplemented his income in this way; his prestige was so great that he did not mind being looked at askance by his colleagues because of his business interests. The patenting of discoveries of commercial value made in university laboratories was also generally discouraged. Penicillin, the outcome of the work started in St. Mary's Hospital, London by Alexander Fleming and pursued in Oxford by Florey and Chain, was not patented; nor was stilboestrol, the synthetic oestrogen which was the product of team work between Charles Dodds at the Middlesex Hospital and Robert Robinson in Oxford. These two products alone have been the mainstay of industries that have made billions of pounds and dollars both for the UK and for other countries. When after the Second World War ICI, the giant of British manufacturing industry, decided to stimulate scientific research and discovery in universities, it did so by instituting a programme of research fellowships, rather than by inviting university scientists to concern themselves directly with ICI's development problems. Not until a few years ago did the Vice-Chancellors' Committee and the University Grants Committee bless the practice of university teachers seeking and accepting industrial consultancies.

The success of the Wolfson scheme can be measured by the fact that in recent years, the University Grants Committee and the Science Research Council, as well as other bodies, have also started to tackle the problem of the university/industry gap by direct measures, and by the fact that in purely monetary terms the Foundation's Scheme has overall generated for the university departments concerned far more than has been spent on it. University teachers have been encouraged to forge links with industry and, correspondingly, undergraduate and postgraduate students have had their eyes opened to the promise that lies in an industrial career.

In addition to providing a record of what has been done, I have tried to point to such lessons as can be derived from the successes and failures of the Scheme. I have followed the progress of most of the projects that have been supported, where possible consulting not only the people in charge, but also their staffs and their industrial collaborators. Of the 126 grants made in the period 1968-80, 102 have been studied in detail. This book includes comments on about half of these. Everyone consulted has shown a high regard for the Wolfson Scheme and has been most courteous and helpful. It has not been possible to thank everyone for this cooperation, but it is much appreciated.

Sir Leonard Wolfson, Chairman of the Wolfson Trustees, has shown a personal interest in the functioning of the scheme, and has allowed full access to the Foundation's records. Mr W. B. H. Lord spent a great deal of time studying the records and in preparing much of the material which comprises Chapters 1 and 2. His help was invaluable.

I also have to thank Professor C. R. Tottle and Mr David Fishlock for their help in the initial stages of assembling the material, and Dr Fergus Allen for help in preparing the material for publication.

Most of all, Lord Zuckerman's contribution should be acknowledged. He originated the Scheme, which it has been his responsibility to oversee on behalf of the Trustees, and has given wise advice throughout the preparation of this book. He has added his own commentary in a Foreword and Postscript.

Warminster
March 1984   L.R.

ABBREVIATIONS

ABRC	Advisory Board for the Research Councils
ACARD	Advisory Council for Applied Research and Development
ACSP	Advisory Council on Scientific Policy
AGR	Advanced Gas-cooled Reactor
BTG	British Technology Group
CAD	Computer-aided Design
CATS	Colleges of Advanced Technology
CBI	Confederation of British Industry
CEGB	Central Electricity Generating Board
CNAA	Council for National Academic Awards
CSP	Council on Scientific Policy
CTL	City Technology Ltd. (City University)
DES	Department of Education and Science
DOI	Department of Industry
DSIR	Department of Scientific and Industrial Research
EPIC	Education in Partnership with Industry or Commerce
ERA	Electrical Research Association
GDP	Gross Domestic Product
MIT	Massachusetts Institute of Technology
MOD	Ministry of Defence
MinTech	Ministry of Technology
NCB	National Coal Board
NEB	National Enterprise Board
NRDC	National Research Development Corporation
PERA	Production Engineering Research Association
QMC	Queen Mary College (University of London)
QMIRL	Queen Mary College Industrial Research Ltd.
SERC	Science and Engineering Research Council
SISTERS	Special Institutions for Scientific and Technological Education and Research
SRC	Science Research Council
SUIC	Salford University Industrial Centre Ltd.
UGC	University Grants Committee
UKAEA	United Kingdom Atomic Energy Authority
UMIST	University of Manchester Institute of Science and Technology
UWIST	University of Wales Institute of Science and Technology

CONTENTS

1

THE BACKGROUND TO THE WOLFSON
TECHNOLOGICAL PROJECTS SCHEME

ORIGINS

The Wolfson Foundation was established in 1955 as a charitable body with funds provided by Sir Isaac Wolfson and his family. During the twenty-five years up to 1980, the Foundation allocated more than £45 million in the furtherance of its purposes, as set out in its Trust Deed. One was the support of institutions of higher education, and by 1965, that is to say, within ten years, about £5 000 000 had been allocated to universities alone. Wolfson College had been founded in Oxford, and a large grant had been made to Churchill College in Cambridge. Grants were given to other colleges in these two ancient institutions, as well as to other universities, to help build halls of residence. Several professorial chairs were endowed, and new laboratories founded—among them metallurgy at Oxford, and mircrobiology at Imperial College, London.

In a new departure, the Trustees decided in 1967 to support university work in fields of applied science which were 'most likely to improve the economic position of Britain and to help the modernization of British industry'. In taking this step, the Foundation was fully aware of the traditional leaning of research workers in universities towards purely academic problems, and was deliberately seeking to encourage the few who believed that they had ideas which were of commercial value to industry. This change in the Foundation's activities was made after careful consideration. The way in which higher education should be supported had been under discussion over the previous three years, with Sir Solly Zuckerman (as he then was), one of the Trustees, urging the Foundation to address itself to what he saw as one of the major problems of the time: the lack of as useful a link between industry and the universities as was potentially there for the making. Whilst British universities enjoyed a deservedly high reputation for the scientific research they fostered, for a variety of reasons British industry did not seem to reap the benefits that it should, and was less successful than others in using science for industrial and commercial gain. As Sir Solly put it, 'We have the ideas, we should share the profits'. It was with the aim of helping to remedy this situation that the Wolfson Technological Projects Scheme was launched, to finance university work that would be suitable for industrial application (preferably in collaboration with local industry), and in this way to promote the exchange of ideas between universities and industry. At the time, the Scheme was a remarkable and innovative step. There had been much talk about the problem, but almost no effective action.

To appreciate the significance of the scheme and its place in the national pattern of technological endeavour, it is useful to consider briefly the history of the relationship of universities to industry and commerce, and in particular to recall the environment of the 1960s—a decade of doubt, uncertainty, and change in higher education, in technology, in industry, and indeed in the nation as a whole.

BRITAIN'S INDUSTRIAL GROWTH AND DECLINE

In its heyday, that is to say before other countries became industrialized, British industry had flourished under the direction of men who as a rule were not university graduates. They were men who had started work in the drawing office or on the shop floor; who had taken their National and Higher National Certificates; who passed the examinations of an appropriate professional institution by part-time study; and who had gained their practical experience by working under the surveillance and guidance of more senior engineers. To them the purpose of research and development was utilitarian: to produce a new or better product, or a new manufacturing process. A commercial attitude was vital in order to achieve sales, to make profits, to pay dividends, and to plough back for future investment.

With the growth of scientific knowledge in the years between the world wars, and as foreign competition for markets increased, industrial firms, especially in the chemical and electrical sectors, began to recruit more and more science and engineering graduates. In the majority of cases they came from the provincial red-brick universities, with which directors of many local industries kept contact. But within industry, there still remained a gulf between those who had a degree and those who had 'come up the hard way'. Lord Hives, the Managing Director of Rolls Royce in its great days, used proudly to declare his preference for the latter. Because such attitudes and differences prevailed, it was generally believed that as a rule only 'lower-quality' graduates went into manufacturing industry, with the best staying on in academic life, and those in the middle range going to governmental research establishments or into 'research centres' which some of the larger companies were forming (significantly enough, often sited well away from their own manufacturing plants).

The two world-wars—particularly the second—did much to bring the two sides together. With all the scientific resources of the nation mobilized for the war effort, university scientists of all disciplines were posted to industrial jobs, often developing new equipment or processes that were urgently needed by the armed forces. Academic scientists also took the lead in developing operational research and other analytical procedures which were needed in the taking of major decisions. This wartime co-operation helped enormously in bringing industry and the universities closer than they had been before, and in eliminating the pre-war prejudices the two had tradition-ally entertained about each other.

Britain emerged from the last war with an optimistic view of the future and, at the same time, inspired with the need for reform. The public had been made aware of the great contributions made to victory by British scien-tists and engineers. Much of the war-winning technology had caught the public eye: new aerodynamic designs; the jet engine; Radar (the American name for the British invention of radio-location); the oil pipeline named Pluto which ran across the Channel; penicillin; and much else.

These technological achievements were responsible for a widespread belief in the immediate post-war years that a prosperous Britain, based on science and technology, could be recreated, with the resurrection of the industrial supremacy which the nation had enjoyed during the nineteenth and early twentieth century. A major difficulty, however, was the country's lack of an adequate supply of scientists and engineers. Without them, there could be no effective recovery in a post-war world which had been, and which was clearly going to continue to be, transformed by the fast-growing power of the United States. To advise on this and other problems related to the administration of science, the Government in 1945 set up the Committee on Future Scientific Policy (the Barlow Committee). One of the most important consequences of its work was the creation of the Advisory Coun-cil on Scientific Policy (the ACSP), a body which soon exercised a major influence on shaping national policy towards science. The ACSP in turn set up a Scientific Manpower Committee, under the chairmanship of Sir Solly Zuckerman, to tackle the important task of increasing the output of scien-tists and engineers from the universities.

A second hindrance to Britain's technological resurgence was the return in many quarters to old habits and ways of thinking, and a lack of apprecia-tion among many who sat on the boards of manufacturing firms of what could be achieved through the vigorous exploitation of science. It was widely but incorrectly assumed that matters would improve if more scientifically-trained people became members of the boards of industrial companies.

Another assumption was that many ideas which were ripe for industrial exploitation were failing to attract support. Because of this, the Govern-ment in 1949 set up the National Research Development Corporation (the NRDC) to help inventors. In effect, the NRDC was a risk-taking agency. It is important to note that there was no 'and' in the title. The Corporation's task was to develop inventions that derived from the results of publicly-funded and other research. Its functions were defined in the Development of Inventions Act 1967, which also required the Corporation 'to attempt to match income with outgoings' taking one year with another. The NRDC had two roles: the first to exploit inventions that were made during the course of Government-sponsored research; the second to help finance innovation in industrial companies.

In the 1960s the NRDC had first claim in practice on ideas from all Government-funded research and development other than the small proportion that was funded by the University Grants Committee (the UGC) in universities. At the beginning of the 1960s, applications were reaching the Corporation from both sources, those from the staff of university laboratories being about a third of those received from members of the government's own establishments. By the end of the decade, the proportions were about equal.

The NRDC did more than simply finance inventions arising in the 'private sector'. It also engaged in partnership ventures with such orthodox financial institutions as merchant banks. It thus provided a means of getting good ideas to the market. The outcome—one-third of the ventures which it supported a complete loss, one-third breaking even, and one-third making a profit—was as good as could be hoped for in the highly speculative and little understood process of successful technological innovation. However, to achieve this success rate, and because it is a public body spending public funds, the Corporation has had to subject the many proposals it has received to so strict a scrutiny that a large proportion of the proposals made to it are rejected. It was inevitable that a certain conservatism, a certain rigidity, was to characterize the Corporation's operations. The widely criticized monopoly power which it used to enjoy over the application of the fruits of government-funded research has recently been abolished by the Government.

In addition to the NRDC, large amounts of taxpayers' money were directed by successive governments into their own research establishments and into government-funded research and development contracts in industry—the larger part in the defence field. Industrial firms were also encouraged to plough back their own money into research and development. As a result, although dwarfed by the United States, Britain soon became one of the heaviest spenders on research and development. Table 1.1 below (based on OECD figures) shows the relative situation in 1963-64, four years before the Wolfson Scheme was launched.

Unfortunately, the returns for all that the Government had been doing were disappointing. Britain's position as an industrial nation continued to decline in the face of competition. There were recurrent crises in the balance of payments, and within industry itself there was an increasing concern at the so-called 'technological gap' which was developing between Britain and, in particular, the United States, but also those industries on the Continent that were rapidly reinvesting. By the late 1950s and early 1960s, it was becoming clear that something was badly wrong in many of our technology-

Table 1.1

Research and development resources in 1963-64

Gross national expenditure on research and development (GERD)

Country	Total GERD (US $ million)	Per head of population (US $ million)	% of GNP	Qualified scientists, engineers and technicians employed on research and development
United States	21 075	110.5	3.7	696 500
Britain	2160	39.8	2.6	159 540
Germany	1436	24.6	1.6	105 010
France	1299	27.1	1.9	85 430
Japan	892	9.3	1.5	187 080

Note: The use of official exhange rates tends to exaggerate the difference between the United States and other countries in the first two columns of figures. (Table from H. W. Pout (1971). Defence research under pressure. In *The new scientist* (ed. D. Fishlock). Oxford University Press.)

based manufacturing companies. The post-war euphoria was giving way to disillusion as it became apparent that the pursuit of science and technology in isolation would not lead automatically to material prosperity.

EDUCATIONAL ATTITUDES AND TECHNOLOGICAL EDUCATION IN THE 1960s

A few scientific subjects, in particular astronomy, mathematics, and medicine, had been taught in the medieval universities from very early times. In general, however, 'science' did not find a firm place in the university curriculum until the latter half of the nineteenth century, particularly in the period when the 'red brick' universities and the Victorian technical colleges were founded. Even then, in accordance with the scholastic tradition which still dominated much university thinking, science was usually treated as an aspect of knowledge that had to be pursued for 'its own sake'. Indeed, it was not until after the First World War that science became fully established as a university discipline, but with engineering still struggling to gain a foothold. With only a few exceptions, for example in chemistry and electrical engineering, where many links existed between university training and industrial applications, the teaching of science in university departments, certainly up to and including the 1960s, was generally carried out by academics with little if any industrial experience, and with scant consideration for the needs of industry.

Throughout the 1950s and early 1960s, the Scientific Manpower Committee, in co-operation with the Treasury and the University Grants Committee, worked to encourage the output of more university-trained engineers and scientists. But the continuing decline of British manufacturing industry in the face of overseas competition, which became more and more apparent in the late 1950s, led to serious questioning of the technological education system and to a call for a revision of attitudes on the part of both the educators and the employers of the educated. The Government then decided to commission a review of the whole higher education system. In February 1961 a committee was set up under the chairmanship of Lord Robbins 'to review the pattern of full-time education in Great Britain and in the light of national needs and resources to advise Her Majesty's Government on what principles its long-term development should be based'. This Report,[1] which was submitted in the autumn of 1963, provided the first comprehensive survey of higher education in this country, the aims of which, as set out by the Committee, provide an interesting insight into the thinking that was current at the time. They were:

> Instruction in skills suitable to play a part in the general division of labour;
> that what is taught should be taught in such a way as to promote the general powers of the mind;
> the advancement of learning;
> and finally
> the transmission of a common culture and common standards of citizenship.

The first of these aims meant instruction in the vocational skills needed to equip people to earn a living, an objective which was placed first on the list, not because it was the most important, but because it is sometimes 'ignored and undervalued'. As the Committee put it, quoting from Confucius's *Analects*, 'it is not easy to find a man who has studied for three years without aiming at pay'. Attention (and respect) has always been paid to vocational needs in some academic fields; for example, people take a degree in medicine because they hope to become physicians or surgeons. But, reflecting the views of many in the academic community in the 1960s, the Committee felt that the prime purpose of higher education must be 'to develop the mind', to advance learning and to ensure the continuity of our culture. Even in the technical colleges, whose most important task is the teaching of vocational skills, there was an underlying belief that learning for its own sake was in some way superior to study for practical purposes.

The Robbins Committee's main recommendation was that the university system should be expanded at a rate which would double the number of full-time students in higher education in a period of ten years. To this end it was advised that especially generous grants should be made to universities other than Oxford and Cambridge in order to help increase their attractiveness relative to the two ancient seats of learning. In effect, the idea was that a university education should cease to be a privilege enjoyed by a minority but should be open to all who could meet the standards set for entry. This recommendation was immediately accepted, thus averting the danger of it becoming a major contentious issue in an electoral year.

Another recommendation was that there should be five institutions (based on existing colleges) to be called Special Institutions for Scientific and Technological Education and Research (SISTERS), the purpose of which, in the words of the Report, was 'to give to technology the prominence that the economic needs of the future will surely demand'. In making this suggestion, the Committee quoted as possible examples or models the Massachusetts Institute of Technology and the Technical High Schools at Zurich and Delft. The Government's initial response to the proposal to establish technological universities was favourable, but in May 1964, the University Grants Committee advised that while it 'had no doubt whatever about the great national importance of expanding and improving technological education', it 'saw serious disadvantages in the concept of SISTERS as a separate and special category of institution'. Its main objection to this recommendation was that 'it would tend to stifle growth elsewhere and draw strength away from those University institutions which already had highly developed faculties of science and engineering'[2]. This advice was accepted, the Government declaring that it was opposed to the idea of giving any institution a special technological designation and that instead it would 'prefer to encourage and expand the many promising developments in the technological departments of other universities'.

Following this decision, the UGC invited Imperial College, London, Manchester College of Science and Technology, and the Royal College of Science and Technology, Glasgow (named in the Robbins Report as three possible SISTERS), to submit proposals on 'the technological developments which they could usefully and practically carry forward in the last two years of the 1962-1967 quinquennium'. These proposals, when received, were considered by the UGC on the basis of three criteria set out in para. 227 of *University Development 1962-1967*. Grants were accordingly made to three selected institutions, and also to the Cranfield Institute of Technology. It is interesting that the UGC criteria did not use the word 'industry', nor did they in any way imply that relevance to British industry would be a factor in considering the merits of the proposals which the universities had been invited to put forward. In retrospect it seems regrettable that the action taken in implementing this part of the recommendations of the Robbins Committee was not more imaginative.

The Robbins Committee also recommended that the Colleges of Advanced Technology (CATS), which had been formed a few years previously from local technical colleges, should be given university status with the power to grant CNAA degrees. This recommendation was also accepted, the Committee warning against the amalgamation of the CATS with existing universities, as this might 'lessen the present predominance of technology'. The Committee also expressed the hope that all higher educational establishments would encourage technological as well as scientific research, and also close co-operation with industry and with governmental research establishments. It is of some significance that the form of this co-operation was envisaged in very general terms, for example, through the exchange of staff between universities and industry (including part-time teaching by staff employed in other fields) rather than in the launching of joint projects, or in the kind of industrially orientated university research later to be encouraged by the Wolfson scheme.

As a direct result of the Robbins Committee's recommendation, the 1960s witnessed a vast expansion in the facilities for higher education. But although the numbers of graduates in science, engineering, and technology increased sharply, there was none the less a drift away from science when measured in relation to the growth of the population of full-time students as a whole. Thus, by 1968, five years after the publication of the Robbins Report, there were twice as many students studying social science as in 1963, whereas the increase in science and technology was only about 1.6 times the 1963 figure. Moreover, not all the new technological universities expanded in the manner hoped for, and one observer of the academic scene was moved to say[3] that 'institutions of the first rank in the technological field have tended to spawn social science and arts faculties instead of pursuing excellence in the manner of MIT or the German Technical High Schools, in order to justify the name "university"'.

The increase in the number of graduates trained in technology, although significant, failed to reach the level anticipated in the plans formulated in the 1960s. On the other hand, many of the teaching institutions, especially the new technological universities, as well as the newly created polytechnics, did start to consider the actual needs of industry and to plan courses to meet those needs. Underlying the whole system of higher education, however, remained the principle of free choice of subject by the student. Universities provided courses in response to student demand, the nature of the demand having been determined during the later years of school life, often influenced by schoolteachers who themselves had no experience of industry or commerce, and who indeed sometimes showed little respect for the new technological institutions.

GOVERNMENT SUPPORT FOR TECHNOLOGY

The national importance of science also became reflected in the machinery of government. In 1959, the Prime Minister, Harold Macmillan, announced that Lord Hailsham, then the Lord President of the Council, would sit in the Cabinet as Minister for Science and Technology, taking over the general responsibility for the Research Councils, which up to then had come under the loose and purely formal aegis of his previous office. He was also to assume responsibility for atomic energy, a task previously shouldered by the Prime Minister himself. In 1964, Harold Wilson, the newly elected Prime Minister, went a step further by creating a Ministry of Technology, under Mr Frank Cousins, with general responsibility for guiding and stimulating a major national effort to bring advanced technology and new processes into British industry. Lord Snow, the scientist/novelist, was appointed spokesman for the new Ministry in the House of Lords, while Professor P. M. S. Blackett (later Lord Blackett) was brought in as Scientific Adviser. But although an Advisory Council on Technology, which was appointed by Mr Cousins, did include industrialists among its members, no full-time industrialist was recruited to his staff. The basic remit of MinTech, as it quickly came to be called, rapidly widened. In addition to a general responsibility for manufacturing industry, it took over from the Board of Trade special responsibility for the machine-tool, computer, electronics, and telecommunications industries. The idea was that MinTech would undertake technical and economic studies for these and other industries in order to show how to stimulate their progress. It was also hoped that the Department would follow up these studies with specific programmes aimed at fostering new technological developments. The Minister was also soon made responsible for the United Kingdom Atomic Energy Authority; for the NRDC; for the support of nearly fifty industrial Research Associations (taken over from the old Department of Scientific and Industrial Research (DSIR), which had been founded in 1916 and which was then dismembered); and for the running of several research establishments.

After the re-election of the Labour Government in 1966, the Ministry of Technology was even further enlarged by being merged with part of the old Ministry of Aviation. By then industrial liaison officers had been appointed in order to facilitate the transfer into industry of the results of university and government-sponsored research. The University Grants Committee also provided a few small pump-priming grants to establish university units to sell technical services to industry. These schemes were not particularly successful. One criticism was that they tended to treat technology as though it were a commodity like a packet of tea, to be bought off the shelf in a corner shop. The technical services units sponsored by the UGC also frequently ended up with high-grade university technologists merely doing semi-routine testing. Neither scheme succeeded in developing industry-university partnerships dedicated to the achievement of mutually agreed objectives.

When MinTech was formed, the small office of the Minister of Science was joined to the existing Ministry of Education to form the Department of Education and Science (DES). DES was responsible for funding education in general, and for the Research Councils. An advisory Council for Scientific Policy (CSP) was then set up to advise on 'the formation and execution' of government science policy. But the CSP, which was the lineal successor of the Advisory Council for Scientific Policy that had been established in 1946, chiefly concerned itself with the funding of the country's five Research Councils. Two of these—the Medical Research Council and the Agricultural Research Council—had been in existence since early in the inter-war years. The other three were created in 1965: the Natural Environment Research Council, the Social Sciences Research Council (now the Economic and Social Research Council), and the Science Research Council (the SRC) (now the Science and Engineering Research Council), the latter being the rump of the old DSIR. The SRC remained responsible for the Royal Observatories at Greenwich and Edinburgh, and for the National Institute for Research in Nuclear Science, that is to say, for an area of basic science. In the first three years of its existence, the SRC strongly favoured what was called 'big science'—nuclear physics, astronomy (particularly radio astronomy), and space projects. These were the big spenders in fundamental research, and accounted for most of its disbursements. For example, in 1965-66 the Council's £28 million budget was spent as follows:[4] nuclear physics, 46.1 per cent; astronomy, space, and radio, 20.6 per cent; non-nuclear physics, 6.3 per cent; biology and human sciences, 5.0 per cent; chemistry, 7.6 per cent; mathematics and computing science, 5.3 per cent; engineering, 5.8 per cent.

The bias towards 'big science' was further emphasized by the organization of the new Council. Three Boards managed its programme. The first dealt with nuclear physics, and the second with space and radio astronomy. The third, the University Science and Technology Board, covered all other fields of science but was responsible for no more than a quarter of all the Council's outlays. In 1969 it was divided into a Science Board and an Engineering Board. Then in 1970 the SRC announced that it was going to 'give continued priority to engineering and the applied sciences'. The job of the Engineering Board was not only the furtherance of knowledge, but also 'its creative deployment in the best interests of society at large'[5]—terms which reflected the aims of higher education as seen by the Robbins Committee.[1]

The Engineering Board of the SRC thus dealt almost entirely with engineering science in the universities. Its support for research was based on the academic criteria of 'excellence, timeliness, and promise'. Work of direct industrial value was not supported. Indeed, as recently as 1975, the SRC declared that if 'the main purpose' of a proposed piece of research 'is specifically to develop a machine or process . . . then it is not normally appropriate for SRC to fund the project'.[6]

Funded as it was by DES, the SRC was remote from MinTech, and although much of the science—'big' and 'small'—which the Council supported was relevant to the objectives of MinTech, there was very little communication between the two bodies. Indeed, there was no incentive for the SRC to orientate its programmes or its thinking towards the needs of British industry.

It was publicly accountable, and was never intended to be a risk-taking body. Not until the end of the 1970s did the Council start supporting research which was directly concerned with industrial problems. By that time attitudes about the purity of academic research had undergone a major change.

With the winding-up of the Board of Trade in 1970, the Ministry of Technology became the Department of Trade and Industry, in the process shedding some of the functions for which MinTech had been made responsible. MinTech had been a main provider of Government funds to help industrially-orientated technology. A considerable amount of Government money was also channelled into advanced technological projects through the Ministry of Aviation (the latter became one of MinTech's constituent parts), and through the Ministry of Defence, which was reconstituted in 1964 by the integration of the three separate Service Ministries. Both Defence and Aviation spent large amounts of money not only in their own in-house research establishments, but also in the support of development projects in industry. In theory, a good deal of the advanced technology which they supported should have had beneficial effects ('spin-off') in civil industry, but despite efforts by the Ministries concerned the amount of spin-off was disappointingly small. This was partly due to considerations of military security, but also because the industrial firms concerned tended to keep their defence activities separate—often geographically separate—from their non-defence work. There was thus a barrier to the transfer of defence-generated technology into civil industrial use.

In 1965 the Confederation of British Industry (CBI) and the Committee of Vice-Chancellors and Principals of the Universities of the United Kingdom organized a joint conference. As the Chairman of the CBI Sir Maurice Laing put it in his opening address, its purpose was to discuss 'the need for change in industry and in education if we are to keep pace with our foreign competitors'. This conference provided the opportunity for exchanges of ideas at a high level, but conventional university attitudes still persisted, so that one speaker from the floor received a measure of support when he said that 'universities existed for the education of good citizens . . . I object to them being used as cram-shops for industry . . . Academicians should be allowed to develop the education of the cream of the population . . . so as to produce citizens of the highest quality for the country as a whole, not simply men who, regardless of all else, will make bigger profits for industry'.

In spite of all that had been said and done, this was still a commonplace view of the role of the universities. But the fact had to be faced that none of the many governmental measures that had been taken to increase the country's technological strength had yet brought much economic benefit in the face of overseas competition. An example of what might be done needed to be set.

2

THE INITIATION OF THE WOLFSON TECHNOLOGICAL PROJECTS SCHEME

The Technological Projects Scheme was launched in 1968 as a logical extension of the Foundation's work in the field of education. Many millions of pounds had already been donated to university and hospital building. Mr (now Sir) Leonard Wolfson, a founder Trustee, was Chairman, with Sir Isaac, his father, President. The former recognized that more could be done to improve the nation's fortunes than supplementing in a general way governmental disbursements to the universities, medical schools, and hospitals. One special problem of the period—the desirability of increasing the numbers of women in higher education—had already become one theme of the Foundation's work. Consideration was also given to what at the time was held to be a major national problem—the 'brain drain', the migration of young scientists and engineers to universities in the United States, drawn there by the greater opportunities and the higher standard of living. Among the solutions that were considered was the endowment of short-term post-doctoral fellowships to persuade the best researchers to stay on in British universities. Another was to set up a fund to help entice young graduates back to the UK. A third was to establish research schools in the universities, even the formation of special institutes. None of these seemed practicable.

From the start of these discussions, Sir Solly Zuckerman, a relatively new Trustee, and at the time Chief Scientific Adviser to the Government—and from previous and current experience aware of the limitations attendant on any action taken by central government—was pressing the need for direct collaboration between universities and industry. He set out the way he thought this could be done in a discussion paper which was considered by the other Trustees.

This paper contained a forthright statement of the situation in 1966 as Zuckerman saw it. After remarking on the recent growth in the number of graduates in science and engineering, the paper went on to say that 'the basic national problem remains unchanged. Our economic problems have not been solved. Productivity remains low. We are not outstanding in industrial innovation. We are slow relative to our competitors in transforming the discoveries of pure science into the kind of technology which is essential if our modern industry is to succeed in world markets . . . what the country needs is not more pure science but a greater willingness on the part of university graduates to face, and a greater skill in facing, the intellectual challenge of applying the fruits of science to economic ends'.

This paper then went on to outline the proposal which was to become the nub of the Technological Projects Scheme—that the Foundation should finance work in universities that was directly relevant to the requirements of industry, and which would serve to bring universities and industry together in joint endeavours.

After detailed discussion with Trustees and various national authorities, the paper, with little change of substance, was formally accepted by the Trustees at their meeting on 7 June 1967.[7] The main recommendation was that funds should be set aside each year for the next five years, 'primarily for the promotion of those fields of applied science which, in the judgement of the Trustees, are likely to improve the economic position of Britain and help the modernization of British Industry'. The Trustees also decided that whatever total sum of money the Foundation allocated to higher education during the following five years, two-thirds should go to the Technological Projects Scheme.

The introduction of the Technological Projects Scheme coincided with the appointment of Major-General Leakey as the new Director of the Foundation. It was his task to transform Sir Solly Zuckerman's outline of policy into a practical scheme. In fact the definitive Zuckerman paper had suggested in detail how the Scheme should be implemented. The Vice-Chancellors of all British universities were to be asked to invite relevant departments to put forward projects for consideration. Oxford and Cambridge and London were not to receive preferential treatment. The new Scheme was to be put to all comers on an equal footing, the total sum allocated by the Trustees being divided between those proposals judged the best. Trustees agreed to be guided by a panel of assessors, under Sir Solly Zuckerman, which was to select the most promising proposals.

The basis for selection had been set out in Sir Solly's paper, and the intention that only joint projects between universities and industry should be supported had been made plain. Thus, in addition to the overall aim of benefiting British industry, it was proposed that 'projects should be assessed mainly in terms of their likely economic impact, taking account of factors such as size of markets, likely effects of the work on cost, quality, demand and sales . . .'.[7]

Links with industry were explicitly called for, especially the 'closest working relationship with local industries, so that the latter are involved *ab initio*'. In addition, 'The projects which should be put to us should be very specifically linked to real problems of industry and should serve as a natural means of bringing universities and industry together, with a mutual interest in each other's problems, purposes and abilities, as well as helping towards the solution of those problems'.[7] What the Trustees approved was not a 'quick-fix' fire-brigade service to encourage skilled university people to solve one-off problems of local industry, but the development of long-term joint endeavours between industry and teams in universities.

The paper which the Trustees agreed also stated that in addition to backing individual short-term development projects, the Foundation might also, in suitable circumstances, help to found small but permanent specialist units to develop and exploit some particular branch of technology for the benefit of industry, with a wide interpretation being placed on the term 'technology'. Assessors were invited to consider projects 'which lie outside the field of orthodox scientific research, but which are concerned with the study, at an advanced level, of industrial methods, processes and problems'.

THE SCHEME IN OPERATION—THE FIRST ROUND

The Foundation's decision to proceed with the Scheme was made public in October 1967. By February 1968, almost all University Vice-Chancellors had shown some interest, and by the closing date of 1 July 1968 about 150 applications had been submitted, chiefly from universities and colleges of technology, but with a few from commercial firms and private individuals. Eleven came from academic groups in Oxford and Cambridge, the eight from Cambridge being the most from a single university. The older civic universities were also well represented. One of the newest universities, Salford, submitted three proposals. There were two from the Hornsey College of Art.

Not all the applications reflected the purposes of the Scheme, and at a first sift seventy proposals were rejected as being far outside its intended scope. The remaining eighty covered a wide range of topics, some of which were thinly-disguised proposals for traditional academic research, with a sentence or two added about possible technological exploitation. In addition to the many proposals with direct or obvious industrial application, others included: elucidation of the problems of land-resource allocation; an enquiry into the institutional demand for tableware in Britain; compilation of a geochemical atlas of the British Isles; a new bleaching method for branding animals; a system for stimulating rainfall in arid regions; and an investigation into the technical and economic impact of a university on its environment.

Sir Solly Zuckerman then convened an expert panel of scientists and technologists to assess and grade the applications and to submit recommendations to the Trustees. Six of the assessors were well-known academics who were sympathetic with the aim of instituting joint projects to strengthen the links between universities and industry. It was agreed that the criteria that would be used in judging the merits of the applications were:

1. That applicants should be known for their intellectual quality.

2. That projects should preferably be associated with some local industry, and should carry a reasonable promise of economic return.

3. That the projects would not normally be eligible for a grant from any other body. It was the Foundation's intention to act as a catalyst rather than compete or conflict with the work of other organizations.

Of the eighty proposals which remained after the first sifting, thirty-five were soon rejected, chiefly because of doubts about the industrial value of what was proposed. The final forty-five were closely studied over the next three months. The assessors discussed the proposals with experts in industry and universities, and also sought the opinions of each of three referees named in the applications. Each assessor, acting as an experienced and critical selector, thus finally arrived at his own judgement of the merits of a proposal.

After the panel's final meeting, eighteen projects were recommended to the Trustees for support, with a further six in reserve. Three of the selected eighteen had a special medical context, and the Trustees decided that these should be supported from Foundation funds not allocated to the Scheme. As a result, the first round resulted in the support of fifteen projects, at a total cost of just over £1 million. Grants were to be payable over a period of up to three years.

Each successful applicant then received a letter from the Director of the Foundation indicating the size of the grant to be made, making it clear that this was a once-for-all payment for the particular project, and requiring the applicant to submit a progress report in six months' time, to report thereafter once a year, and to discuss any possible patent rights and commercial applications.

These conditions represented a balance between research and business needs. The call to make annual reports to the Trustees was hardly onerous, but the need to provide the first report six months after receiving the grant imparted a sense of urgency to the work. In later years, these annual reports were supplemented by visits of representatives of the Foundation to the laboratories concerned. The other stipulation—that the grant was to be a once-for-all payment—was made in order to emphasize that the Foundation was not intending to support open-ended research projects, but was providing an opportunity of applying existing knowledge to achieve a practical outcome.

The Foundation also reserved the right (if it chose to take it) of gaining some reimbursement from the more successful ventures, with the intention of recycling such money into new projects.

The million pounds allocation to the first round of the scheme supported a wide variety of projects, but as a broad generalization the funding took one of two forms:

Either to support the development of a specific idea for a set period, up to a point when it was expected to be commercially viable;

Or as 'pump-priming' money to set up a unit which would provide services to industry on a repayment basis and hence, if successful, become self-supporting.

There are examples of success and failure in both these categories, and outlines of individual cases are given in Chapters 3 and 4.

THE SCHEME IN OPERATION—LATER ROUNDS

The sequence of grants made in later years closely followed the pattern used at the start. A second round in 1970 supported twelve projects in eight universities at a total cost of £741 730. These twelve were chosen from over seventy applications. In retrospect, there is no doubt that a number of very good projects failed to gain support because the funds needed were not available. In the third round of the scheme in 1972 sixteen projects were supported at a cost of £1 230 592.

In the fourth, proposals were specifically invited for projects relating to the development, or the better use, of the country's own natural resources. Some 150 applications were received, and eighteen were selected for support at a cost of £1 009 003. In 1976 twenty-three proposals costing a total of £2 429 900 were accepted; in 1978 twenty-one costing £1 931 070; in 1980, twenty-one costing £2 000 000; in 1981, twenty-four costing £2 255 000; in 1982/83, seventeen costing £2 130 000, and in 1983/84 a further nineteen grants costing about £2 000 000 have been made.

From its inception the Scheme has supported 186 projects at a total cost of £17 million.

THE SCHEME AS A PACE-SETTER

A constant feature of the Scheme has been the very large number of applications submitted in each round. There is clearly no shortage of practical ideas in the universities. As a private charity, and moreover one with very wide interests, the Wolfson Foundation has clearly been limited in the amount of money it could make available. The scheme was commended in the scientific press as being highly imaginative and, starting when it did, it not only aroused wide interest, but also focused attention in a very practical way on an issue about which there had been much discussion but little action. The Foundation can truly claim that it has blazed a trail, and one which Government-funded agencies were later to follow. Whatever the commercial value of the projects that have been supported, or the benefit to industry of stimulating the interest of university departments in their work, the Foundation from the start provided a working model of a direct way of encouraging industrial and academic collaboration.

3

PRODUCT AND PROCESS INNOVATION

The titles of the diverse projects that have been supported under the Wolfson scheme, and the university departments that have benefited, are listed in the Appendix (p. 89). The majority are classified in Tables 3.1 and 3.2 under eight broad subject headings and in accordance with seven categories of aims as stated in the applications. (The awards made in 1981, 1982, and 1983/84 are not included in the Tables, since it is too soon to make an assessment of the results that have been achieved.)

Subject headings	Aims
Agriculture and fisheries	To devise a new product
Instruments	To improve an existing product
Mechanical, production, and civil engineering; robotics	To devise a new process
Electrical, electronic, and software engineering	To provide research and development service
Metallurgy and materials technology	To establish a research or testing facility
Energy	To establish a new institute or unit
Chemistry and chemical engineering	
Biochemistry and biotechnology	

Of the fifty-six possible combinations of subject and aim—in the sense that a project under one of the eight subject headings might have indicated as its purpose any one of the seven aims that are listed—Tables 3.1 and 3.2 contain entries that account for forty-five of the fifty-six combinations, with some three grants on average in each.

Even with so many possible combinations of subject and aim, the decision about the classification of some projects has had to be somewhat arbitrary. For example, the subject-heading or category 'electrical, electronic, and software engineering' embraces, at one extreme, the thermal imaging of miniature integrated-circuit devices (Table 3.2, Birmingham, 1980) and, at

Table 3.1
Products and processes

Subject group	Aim or achievement		
	New product	Improved product	New process
Agriculture, fisheries, etc.	Aberdeen, 1970; £17 500	Aberdeen, 1980B; £61 000 Aberystwyth, 1974; £14 000	Birmingham, 1978A; £164 000
Instruments	Imperial College, 1978B; £75 000 York, 1978; £100 000	Aberdeen, 1980A; £50 000	Manchester, 1976; £156 000
Mechanical, production, and civil engineering robotics	Aberdeen, 1976; £72 000 Cranfield, 1980; £52 000 Queen Mary College, 1976; £238 000	Bath, 1972; £92 000 Bath, 1978B; £160 000 Bath, 1978C; £50 000 Manchester, 1978; £184 000	Aston, 1976; £100 000 Birmingham, 1968; £67 000 Nottingham, 1972; £45 000
Electrical, electronic, and software engineering	Bath, 1976; £102 000 Essex, 1980B; £100 000 Queen Mary College, 1974; £93 000 St. Bartholomew's, 1976; £184 000	Newcastle, 1976; £155 000 Strathclyde, 1976; £110 000 Strathclyde, 1978; £107 000	Bangor, 1978; £56 000 Bath, 1980A; £127 000 UMIST, 1978; £45 000 Salford, 1976A; £96 000
Metallurgy and materials technology		Nottingham, 1980; £136 000	Birmingham, 1970; £128 000 Liverpool, 1974B; £62 000 Liverpool, 1976; £120 000 Sheffield, 1978A; £40 000
Energy		UMIST, 1980; £113 000	Belfast, 1972; £54 000 Cranfield, 1974; £155 000 Salford, 1976B; £75 000
Chemistry and chemical engineering	Bath, 1978A; £49 400 City, 1972; £36 000 Edinburgh, 1972; £92 000		
Biochemistry and biotechnology	Bath, 1980B; £100 000 Cambridge, 1980; £156 000 Kent, 1980; £90 000 Ulster, 1978; £141 000	Strathclyde, 1980; £88 000	Imperial College, 1976; £46 000 Manchester, 1974; £68 000 Reading, 1974A; £120 000 Sheffield, 1978B; £120 000 York, 1974; £44 000

Table 3.2
Services

Subject group	Aim or achievement			
	Research or development	Advisory service	Research or testing facility	Institute or university centre
Agriculture, fisheries, etc.	Belfast, 1978; £49 500 Liverpool, 1974A; £6500 Reading, 1974B; £129 000	Southampton, 1968A; £24 000 Stirling, 1976; £68 000 Stirling, 1978; £38 000		Southampton, 1968B; £30 000
Instruments	Cambridge, 1968B; £50 000 Surrey, 1968; £132 000		Belfast, 1970; £75 000	
Mechanical, production, and civil engineering robotics	Newcastle, 1970B; £77 000 Dundee, 1972; £99 000 Salford, 1980; £105 000	Manchester, 1980; £190 000		Heriot Watt, 1972; £178 000 Southampton, 1970C; £16 500 Birmingham, 1978B; £120 000
Electrical, electronic, and software engineering	Sussex, 1972; £127 600 Warwick, 1972; £148 000	Belfast, 1976; £53 000 Birmingham, 1978C; £95 000 Essex, 1980C; £60 000 Southampton, 1976A; £26 000 York, 1980; £60 000	Birmingham, 1980; £110 000 Essex, 1972; £33 000 Essex, 1980A; £39 000	Edinburgh, 1968; £131 000 Southampton, 1970B; £25 000 Southampton, 1976B; £48 000
Metallurgy and materials technology	Cambridge, 1972; £39 000 Imperial College, 1972; £73 000 Imperial College, 1974A; £60 000 Imperial College, 1974B; £106 000 Imperial College, 1978A; £125 000 Newcastle, 1970A; £156 000 Newcastle, 1978; £18 000 UWIST, 1968; £132 000	Aston, 1970; £17 600 Aston, 1972; £75 000		Nottingham, 1968; £255 000 Southampton, 1970A; £50 000
Energy	Dundee, 1978; £35 000 Heriot Watt, 1972; £178 000 Newcastle, 1972; £38 000			Cardiff, 1980; £54 000
Chemistry and chemical engineering	Leeds, 1972; £84 000 Imperial College, 1968A; £47 000	Surrey, 1974; £39 000		Southampton, 1978; £163 000
Biochemistry and biotechnology	Cardiff, 1970B; £40 000 Strathclyde, 1974; £28 000		Ulster, 1980; £70 000	Southampton, 1976D; £136 000

the other, the puffer-switchgear project (Table 3.1, Strathclyde, 1976), the aim of which was to develop a means of interrupting electrical currents of the magnitude of 20 kA. The former project is concerned in the first instance with research or testing facilities, the latter with a new product. Software development for computers might have merited a group of its own, but is included under the same general heading of electrical and electronic engineering. Simplifications cannot be avoided if comparisons are to be made.

In the following pages the results of a few of the projects are described to illustrate the variety of the work that has been carried out by the University Departments that have been supported. In the seven-part classification of aims, three relate to the development of *new products or new processes*, and are included in Table 3.1, and Table 3.2 lists four concerning the provision of *services to industry*. A comparison of the listing of the tables with the titles given in the Appendix indicates that changes have sometimes occurred as work on a project has proceeded, with a shift of emphasis from one category to another. A feature of the administration of the Scheme has always been an absence of formality in dealings between the Foundation and the units or institutes which it has helped to launch, so that agreement can be reached quickly when circumstances call for a change in the direction of the work that has been supported.

New or improved products or processes (Table 3.1) are considered first. The cumulative sum awarded for this category of projects accounted for 52 per cent of the total for all grants specified in Tables 3.1 and 3.2. The full title and size of grant of each of the projects referred to under these headings will be found in the Appendix (p. 89).

NEW OR IMPROVED PRODUCTS

Instrument development, mechanical engineering, electrical engineering, and biochemistry predominate in this section.

Instrument development

Edinburgh University, Department of Chemistry,
1972, £92 500: High-speed chromatograph

The declared aim of the earliest grant, which was awarded to the Chemistry Department of Edinburgh University, was to develop a high-speed chromatograph. The work proceeded to the stage where prototypes of the instrument were constructed, and a new material called Hypersil was developed for packing the chromatograph's columns. Further work was carried out for a time in collaboration with a small company, but difficulties in financing and marketing the chromatograph arose, and in the end a separate arrangement was made for the production and marketing of Hypersil. Sales of this material, plus royalty payments, have allowed the unit to continue on a modest scale, but it is hard to forecast any growth or expansion in the near future. Enough is now earned annually to maintain the unit in being, the total earned so far exceeding the Foundation's initial grant by a considerable margin. The problem of selecting a suitable company for industrial collaboration is one that has arisen with a number of other projects discussed in this and the following chapter.

York University, Department of Chemistry,
1978, £100 000: Infrared analyser

The originator of this project was an industrial scientist who had set out to design an infrared analysing apparatus for use in the milk and food industries. In 1977 he established his own company, and by February 1978 had produced a prototype instrument for evaluation and approval for milk testing in North America. After a short test period, the instrument was accepted as suitably accurate for the determination of fat, protein, and lactose in supplies of milk from individual cows and, hence, for certification for payment to individual producers.

The designer sought help from the Chemistry Department at York University, which in turn was awarded a Wolfson grant to further the project. As a result it became possible to proceed from the production prototype stage to a saleable design. During the period of the grant, steps were taken to incorporate microprocessors in the design in order to make it easier to handle the information registered by the instrument, an improved version of which will, it is hoped, become available in 1984.

The sequence of events in this project began with the appreciation by an industrial scientist of the existence of a potential market. He then established a close working relationship with a University team, with a division of responsibility that was clear and sensible. The scale of the development programme was within the compass of the small group, and sales soon started to grow. Other new or improved products are expected to emerge.

The company now employs a staff of forty-seven at home and overseas. Its turnover, which is still growing, now exceeds £1.5 million, and the cost of its development programme is about £200 000 a year. In 1982 the company received the Queen's Award for Export Achievement.

An interesting aspect of the Edinburgh and York projects is the fact that the small size of the two companies eased the process of collaboration with university research workers, who were able to play a direct part in the development of marketable products.

Civil and mechanical engineering

While the development and exploitation of instruments may be suitable for small companies, this is not the case when mass production is involved or when manufacture demands the use of heavy equipment.

London University, Queen Mary College, Department of Civil Engineering, 1976, £238 000: Novel anchor design

Queen Mary College's anchor project was based on an invention of members of the Civil Engineering Department. The main features of the new anchor are a top thrust-plate which provides resistance to horizontal movement, and bottom flukes which when open provide vertical restraint, the top and bottom portions being connected by a tube which acts as a friction pile. The anchor can be driven into the sea-bed by a submersible piling hammer and is then held in place to provide permanent mooring.

The early laboratory studies were carried to completion in tests of model anchors under various soil conditions. The data so obtained were used to design small prototype anchors for testing on land and in shallow estuarial waters. This work made possible the construction of anchors for use in different situations and for different purposes, including deep-water offshore oilfield development. A total of twenty-seven anchors, with capacities up to 100 tonnes, have now been successfully installed on a commercial basis for Britoil and British Gas.

The design is suitable for ships of widely different size, given compatible mooring buoys, and is particularly applicable for very large ships in difficult waters. Considerable masses of metal are needed to hold a large ship, and the manufacture of such anchors necessitates partnerships with organizations involved in heavy metal fabrication, and with contractors having experience of offshore engineering.

The difference in cost between a conventional system of anchorage and the Queen Mary College design would appear to be significant. Prospective purchasers are, however, naturally cautious about new designs, and this has retarded market penetration.

Queen Mary College arranged for the finances of the anchor project to be handled by its own company, Queen Mary College Industrial Research Ltd., the company, in association with academic staff, being responsible from the beginning for development and exploitation. Income is expected in the form of royalties, as well as from design and on-site consultancy fees. The first agreement for licensing—now ended—produced a return of £101 000 over a two-year period. A second agreement is currently under negotiation, and plans are in hand for further commercial anchor placements with off-shore contractors.

This pattern of operation, with a university company being set up to carry the responsibility for development work, and to act as agent and negotiator, is clearly a useful model for the handling of any large engineering projects which may be launched by academic departments.

Bath University, Department of Engineering,
1972, £92 000; 1978, £50 000: Diesel engines

Diesel engines provide an example of a field of technical development where exploitation should prove easier than it has so far done for heavy anchorages. Since several engine manufacturers are in the market, any significant advantage in performance or in production costs could improve the market share of whoever adopts an improved design. On the other hand, the increase in sales revenue that might be expected from radical changes in design could well be nullified by the high capital costs which might be entailed.

This was fully appreciated when a grant was made in 1972 to the Engineering School at Bath University for the research stage of a diesel-engine study. In addition to equipment and staff, funds were provided for laboratory buildings, and work now continues as part of the School's research programme.

The grant provided the University with a unique facility for testing both engines and transmissions. This enabled detailed analyses to be made of the steady-state and dynamic behaviour of a wide variety of engine-transmission systems for heavy vehicles. The concept of an integrated approach to the solution of engine-transmission problems was relatively new at the time, since traditionally engines and transmission systems have been studied separately. An integrated approach leads more easily to the development of control and data-acquisition equipment, which can subsequently be used to minimize fuel consumption. Novel types of hydrostatic dynamometers which have been developed during the course of this research are being exploited through a licence granted to a British manufacturer.

Starting in 1974, but mostly from 1977 onwards, industrial contracts to the extent of £400 000 have been negotiated by the engineering school. As a result of this promising start, the University was given a second Wolfson grant of £50 000 in 1978 to help in exploitation. Contracts have not been confined to the field of engine-transmission, but have extended to a wide range of comparable studies, including compound-engine development and microprocessor control for continuous optimization of engine performance.

The Wolfson grants have made it possible to establish a strong research group in the Bath Engineering Department. This in turn has attracted Science and Engineering Research Council (SERC) funds for further basic research, contract income from engine and other manufacturers, and some fees from licensing arrangements with component manufacturers. In money terms, revenue has exceeded the total of the two Wolfson grants by at least three times.

Manchester University, Department of Engineering,
1978, £183 660: New motorcycle design

The work of the Wolfson Motorcycle Unit is directed mainly to highly innovative design and development rather than to research. Initially, the focus was on the design of front-wheel forks, with the idea of improving lateral stability at high speed, especially in heavier motorcycles. The improved stiffness and overall design concept have been validated by trials carried out by overseas manufacturers of prototype units, but attempts to persuade a manufacturer to adopt the new design of fork have not as yet been successful.

The unit then investigated the stiffness of the whole machine-frame, and proceeded to develop a motorcycle engine with better balance. This in turn led to further development of components such as the starter, and the improvement of balance between engine and transmission. Wishing to match good performance with improved safety, the unit then undertook studies of wheel and brake systems which would keep pace with the improved performance of the front forks. Although individual components have to be investigated separately, the Manchester team inevitably became committed to the design of a complete machine. Its four-year programme of work has now ended, but because of the demise of the British motorcycle industry, support has not been found which would permit of the construction of a complete motorcycle. An engine that should satisfy all legislative and type-approval tests was, however, constructed and is now under test. Fuel consumption is expected to be 80 m.p.g. (28 k.p.l.) from a 400 cc single-cylinder engine, with a top speed of 100 m.p.h. (160 k.p.h.). A motorcycle incorporating the Manchester design should be safer, more economical, and less noisy than current machines. Its standards of handling, stability, and overall performance will be very high. Design work already undertaken by the Unit would enable the complete motorcycle concept to be developed very quickly.

In spite of setbacks owing to the lack of major industrial support, the Manchester Motorcycle Unit has become self-supporting. During 1982-83 it established a company within the University to provide a consulting service which now has contracts with European and Japanese manufacturers. Despite the recent downturn in the world motorcycle market, a major European manufacturer is now assessing the feasibility of offering the leading link front forks for a new motorcycle to be announced around 1985.

The story of the Queen Mary College, Bath, and Manchester mechanical engineering projects that have been described above reveals some of the difficulties that are encountered in marketing new products based on the results of university research and development. When the product is physically large and sales are likely to be intermittent, as in the example of the anchor, a university unit will find it difficult to participate at the stage of industrial exploitation. This is also true of mass-produced articles where the product is likely to cost a few thousand pounds per unit. In such cases, the only course then open to a university unit which wishes to become self-supporting is for it to provide a consultancy service to firms in the same or in related fields of manufacture. When the product is relatively small and inexpensive, the university unit could organize itself to cope with questions of design for production and marketing.

Aberdeen University, Department of Engineering, 1976, £71 500: Electrochemical discharge machining

A grant was made to Aberdeen University for the exploitation of electrochemical arc machining, a technology which had been under investigation in the University's Engineering Department for some time, and in the course of which it had been noted that great quantities of metal are removed under certain conditions of discharge. The Wolfson grant was to be used in developing devices based on this observation, particularly when applied to the 'deburring' of metal objects during the course of manufacture. The electrochemical process is able to remove surface irregularities very rapidly without any appreciable loss of the underlying stock metal. The idea was patented by the NRDC, which approached a company with a view to involving it in the manufacture and sales of suitable tools. NRDC support for inventions is usually dependent on a substantial commitment from a collaborating company, and for a variety of reasons this was not forthcoming to the extent required. Although initially NRDC seemed the appropriate agency for patent protection, as it is for an invention arising from Government support, work carried out on a Foundation grant is not necessarily patented, and an amicable arrangement was made for the rights to be re-assigned to the original investigator and then for exploitation rights to be solely held by the company. In the event, the company, which was a subsidiary of a much larger corporation, was forced to retrench as a result of the recession of recent years, and the link with the University was consequently broken.

Negotiations that followed led to the return of the patent rights to the investigator and, after making a survey of possible markets for the machine that had been devised, it was decided to seek other industrial collaborators. During this period the unit's researches were maintained with the help of substantial funds provided by the SERC.

It proved impossible to find an alternative industrial collaborator, and the NRDC was approached once more for help in setting up a new company which would manufacture the electrochemical arc machines that had been developed, but which in addition would produce more conventional electromechanical machines for which a market already existed. It was not intended that the company should be part of the University. Academic staff who were to become involved were to do so, not in their capacity as University employees, but as private individuals. Research staff in the University would, however, continue to provide expert help. The realization of this arrangement was held up by the merger of the NRDC and the National Enterprise

Board to form the British Technology Group (BTG), and during the delay the leader of the University unit and some of his industrial colleagues decided to strike out on their own and form a new company. This was registered in October 1981 and is now trading.

On the appointment of the most senior member of the Unit to the Regius Chair of Engineering in Edinburgh, other members of the team moved with him. The work will now be continued there after the Wolfson grant comes to an end.

Electrical Engineering

Strathclyde University, Department of Electrical Engineering,
1976, £109 500: Electrical switchgear

The problems and hazards of university collaboration with heavy engineering firms have also been demonstrated in the case of a grant to Strathclyde. This concerned a project to develop high-power electrical switchgear. The aim was to develop to prototype stage a high-current interrupter with a rating of 15 kV and a short-circuit current of 25 kA.

To minimize the risk that electrical circuit-breakers may fail because of the formation of arc discharge at the interrupter, the break is either carried out in a bath of insulating fluid or the arc is extinguished by a blast of air as it forms. A third alternative is to use an electronegative gas, such as sulphur hexafluoride (SF_6), to suppress the effects of arcing. The project team proposed to investigate a design concept in which the energy of the arc is utilized to develop the flow of gas. The team was also interested in the use of a piston-operated 'puffer' arrangement, for cases where the energy available from the arc is insufficient to generate the required rate of gas-flow. This gave the project the title 'SF_6-puffer circuit-breaker'.

There is a wide range of possible designs in such a device, involving, for example, the arrangement of contacts, the design of the puffer, and the means of utilizing the energy from the arc to produce a flow of gas. Three test-rigs were designed to provide basic information at the research stage of the project. Further test-rigs and trials were then necessary to optimize all possible variables to meet the performance specification in the best and most economical manner. The actual programme of tests delineated the capabilities and the limitations of the design principles, and the conclusion was that the separate performance requirements could be met.

Many lessons were learned in the course of this particular project. The collaborating company had trouble with its work force, and as a result progress was slowed down for much of the three and a half years of investigation. Indeed, at times the company ceased contributing anything to the test programme—which is not to be wondered at since the commitment of a collaborating company to a joint project is inevitably influenced by market forces and by other considerations affecting company policy. In consequence the University group had to maintain close contact, not only with the technical development department of the company, but also with its design, production, and marketing departments. Had it not done so, it is unlikely that a commercial product could have materialized. The programme of work was so arranged that the University staff carried out the research and development work for which they and their equipment were best suited, while the company supplied other services, with some overlap in the testing of the design concepts. In general the testing of high-power equipment had to be carried out by the firm.

It is unlikely that this project will lead to the immediate production of an assisted SF_6-puffer interrupter, but there is a reasonable chance that a new product will materialize in two to three years' time, and that royalties will then become payable to the University.

As with many other of the Wolfson projects described in this and the next chapter, it is no bad thing if university staff learn to appreciate the problems which face an industrial company in developing a new product from the results of basic research. Correspondingly, it is all to the good when manufacturing firms recognize that high technical and innovative talent can be found within the universities.

Strathclyde University, Department of Electrical Engineering, 1978, £106 170: Small electric motors

Over many years the Electrical Engineering Department at Strathclyde University had developed an expertise in the mathematical modelling and design optimization of small electric motors. For that reason the Foundation made a grant to enable those concerned to collaborate with a manufacturer to produce a novel range of small motors for use in domestic and other appliances. At the start of the programme the respective contributions of the University and the manufacturer were carefully defined in a collaboration agreement. Although there was no call for special high-power test equipment in developing such small devices, the industrial company had a better knowledge of the service requirements and the necessary proving tests. As a result of this collaboration, non-standard prototypes were produced for testing, using facilities made available by the Wolfson grant—facilities not generally available in a mass-production factory. With the theoretical design capability built up in the University, and the production facilities which the grant made possible, priority was given to setting up test equipment so that predictions could be compared with actual performance. The whole exercise was carefully planned and costed against time to produce new prototypes to agreed target dates.

The technical aspects of the programme have progressed satisfactorily, and most of the initial aims have been met. The computational methods used in the University have been extended, sample machines assembled, and an exchange of staff has taken place between the University and the company. New designs, including reversing motors, have been produced as prototypes and have shown that the design methods are satisfactory. There is little doubt that these new designs will result in reductions in manufacturing costs, and in improvements in the efficiency of components. Some work remains to be done, but once again the present recession has affected the manufacturer, with the loss of a high-volume market, and with financial losses resulting from large redundancy payments. It would appear that a reduction of some 20 per cent in manufacturing costs would be necessary to justify re-tooling. The Strathclyde unit is now seeking patent protection in order to safeguard the new knowledge that has been obtained until an alternative outlet is found, or until a satisfactory trading situation arises.

Chemistry and biochemistry

City University, Department of Chemistry,
1972, £36 050: Gas sensors

The Wolfson Unit for Electrochemical Technology at the City University has established a different relationship with industry from the ones just described. This Unit had originally been launched to design an oxygen gas-sensor with the help of a small pump-priming grant from the UGC, and had then set up a joint electrochemical enterprise with the Electrical Research Association as an equal partner. At this point the Wolfson Foundation made a grant of £36 050, which allowed the Unit to expand and develop a service on a customer-contractor basis. One year after the expiry of the grant, the initial objective of becoming self-supporting on contract income had been achieved. Sufficient reserves had also accumulated to allow for the further development of a device which quickly found a market.

The instrument comprises a metal air-cell in which the current output is limited by the rate of arrival of oxygen at one electrode. With the output in a sensitive range, and the rate of diffusion of oxygen from outside the cell restricted by a suitable barrier, a simple measurement can be made of the oxygen concentration outside the cell. The sensors are simple in design and easy to assemble, and the details of the concept are well described in the patent literature. The attractiveness of the device lies in its wide range of application in industry, both for the monitoring of oxygen and the measurement of other gases. It is invaluable in mines and sewers where it is necessary to ensure the safety of the operators by continuous monitoring of the atmosphere; in medical incubators for postnatal care; and for monitoring the oxygen content of the inert gas used to blanket the volatile fluids conveyed in oil tanks and in chemical plants. An equally important use is in the monitoring of flue gas from fossil-fuel boilers to regulate the efficiency of combustion and to economize on fuel.

The University chose to promote its own company, called City Technology Ltd. (CTL), and to give the original research and development team the task of developing the instrument to the stage of production and sale. CTL's efforts were concentrated on the development and manufacture of the sensors. At the same time links were formed with suitable instrument-making companies to find market outlets. The companies had to undertake the design and production of the commercial instruments into which CTL sensors were incorporated.

The small Wolfson grant in the years 1972-75 came at an opportune moment. It enabled the preliminary research and development work to be completed. It also helped in the setting up of the CTL company and in the preliminary stages of manufacture. Towards the end of this initial period, when the company was being established and manufacturing facilities provided, the cash-flow was negative. Such reserves as had been built up kept the company going until sales of the product began to produce an inflow of cash. A further small grant was provided in 1981—one of the few cases where there has been a follow-on grant—to sponsor the development of sensors for the detection of gases other than oxygen.

The company now derives its revenue from the sales of oxygen sensors (at present exceeding £0.5 million per annum), and from the growing sales of carbon monoxide sensors. The instruments into which they are incorporated sell for many times the price CTL charges for the sensor, so that the total new sales arising from the whole development amount to several million pounds a year. This remarkable result is due primarily to the initiative shown by the leaders of the Wolfson Unit.

The decision by the University company to confine itself to production of the sensor and not to become involved in the actual production of instruments or control equipment, proved a wise one. Concentration on production of the sensor alone meant that no heavy or expensive machinery was required. The small components from which the sensor is assembled are obtainable from subcontractors as required. In these circumstances, the working capital required for tools, stores or work in progress, has been kept to a minimum. Only relatively unskilled labour is required, and this is readily available in the neighbourhood of the University. All these conditions are ideal for the formation of a small company, and are in marked contrast to the circumstances which surrounded the Queen Mary College anchor project.

Bath University, School of Chemical Engineering,
1978, £49 400: Protein from bone

In 1978 the Foundation supported a two-year project whose aim was to design a continuous process for the extraction of protein from bone for use in animal feedstuffs. This idea had been conceived as a result of direct contacts between an interested small firm and members of the Chemical Engineering School of Bath University. Studies were first made of the chemical process of leaching bone with acid to reduce the amount of calcium phosphate retained in the bone protein (osein). A flow-sheet of a possible process was produced from these bench-scale studies, and this was followed by a preliminary chemical engineering study. The resulting engineering flow-sheet was tested on a laboratory scale to determine the optimum size of the vessels to be used, of flow-rates, and separation methods. A counter-current plant was then designed to yield an adequate and efficient recovery of the required product from the raw materials.

Close contact was maintained with the collaborating company throughout the university stages of the project. This led to a scaling up of the process vessels to 750-litre capacity. Patents were applied for and, with NRDC help, the manufacturer decided to build a plant to treat 1 tonne of raw bone per hour. The final design of the plant was achieved almost exactly on schedule. The scale-up of the process as a whole from a rig consisting of a series of 5-litre vessels to a plant with 6000-litre vessels progressed satisfactorily. Market trials of the extracted bone protein have proved satisfactory, and further plants are expected to be built as demand increases.

It would have been imposssible to embark on this project without mutual confidence and trust between the university team and the firm. Experience has shown that this is best assured by previous co-operation on other, even small, projects. To offset the financial risks the small firm may well have to seek financial help from an outside body such as the NRDC or some other Government agency. In cases such as these, it is important for patent protection to be sought whenever possible and, given success, for the university to be assured of an adequate share of any profits that are made. In many ways the experience of this project is a model of how university expertise can contribute to the succeeding stages of an industrial development, from the basic science, through the bench-scale and large prototypes, to a plant suitable for market testing.

New University of Ulster, Department of Biological Sciences, 1978, £140 580: Animal feed from distillery waste

A similar feedstuff project which has reached a corresponding stage of development as at Bath is being undertaken by the New University of Ulster, where an attempt is being made to increase the value of a product derived from the highly acid waste, known as pot ale, in whisky distilleries. The material contains quantities of carbohydrate, protein, amino-acids, phosphorus compounds, and glycerol, all of which have potential value as feedstuffs. Part of the distillery waste has customarily been disposed of as an animal feed, known as 'distillers dark grains'. The aim of the University project was to reduce the energy required to produce the feedstuff, and to increase the nutritional value by increasing the protein content. As the project developed, what are known as 'golden grains' were produced with a higher nutritional value than the dark grains.

Pot ale also contains small quantities of dissolved copper, derived from the copper pot-stills, and the conditions under which its presence can be controlled have been identified. This is important for two reasons: first, the presence of copper could inhibit microbial action and, second, it could make the product toxic when used as animal feedstuff.

Before the award of a grant from the Foundation, much research had been carried out into ways of reducing toxicity and of selecting a suitable organism for the production of biomass. The effort was greatly increased following the award, and the design of components for the development stage of the process was undertaken. Early in 1980 there was a proposal for the use of two further micro-organisms (other than the single one originally used), and in 1979 development began on a laboratory scale of vessels for batch or continuous processing of untreated waste liquors.

In this type of process the design of the fermenting vessel is important. The University developed a relatively simple model with a capacity of 100 litres. This proved satisfactory. A 500-litre fermenter was then built and successfully operated. To cope with all the spent liquor that is available from the distillery, a vessel of forty to a hundred times this size would be required. It therefore became necessary to build an intermediate-scale pilot plant of about 1500-litre capacity. This plant was built in the distillery and run during 1982/83 with money provided by the distillery itself, supplemented by a Research and Development grant from the Department of Economic Development, Northern Ireland. This pilot plant made it possible to extend the studies of harvesting and drying the single-cell protein which is produced. The non-toxicity of the product has been tested by feeding it to rats and fish, so far with satisfactory results. Additional feeding trials with farm animals have now been arranged.

The pilot plant of 1500-litre capacity cost about £40 000. A full-scale plant might therefore cost more than £500 000. The unit running costs would, however, be considerably reduced, particularly in fuel, compared with existing methods of treatment, and the output would approach 8 tonnes per day of a pelleted feed-material of about 25 per cent protein content. Both the University team and the co-operating firm are confident that the project will pay off.

More definitely in the biochemical field than either the Ulster distillery-waste project or the Bath work on bone protein is the Cambridge University work on the preservation of fruit.

Cambridge University, Department of Applied Biology,
1980, £156 000: Preservation of fruit

Knowledge derived from studies of the physical chemistry of the surface of fruit suggested that suitable coating of the surface of freshly picked produce could extend the life of certain fruits and vegetables after harvesting, without the use of refrigeration or the provision of special atmospheric conditions. The benefit of such a system would be the elimination of the need to control temperature and atmosphere not only in storage but also during transport, both of which are costly in capital and operation. These conditions are also difficult to apply economically in the early stages after picking and during transportation to seaport or railhead.

The new technique was developed with financial help provided by the Wolfson Foundation, and trials have been made with bananas transported by sea between the West Indies and the UK, and from the Philippines to Hong Kong. No deleterious effects occurred in tests in which a few tonnes of bananas were shipped from the Windward Islands to the UK at temperatures of 22 °C. On the Philippines to Hong Kong route the tests were successful at temperatures of up to 34 °C. The fruit was transported in standard dry-freight containers on deck with the doors open—conditions which would have totally destroyed untreated fruit.

There are other benefits to be gained from treating fruit in this way. Untreated bananas at the edible stage have a shelf-life of only five or six days, which makes it necessary to clear stocks before a weekend in order to avoid decay and waste. On the other hand, the shelf-life of treated bananas is twelve days. Treated fruit can therefore be carried forward for sale into the early part of the following week. Furthermore, fruit treated in the green state can be held in store and brought to the degree of ripeness needed for sale as and when required.

The main benefit, however, is in transportation. Over and above the reduction in cost achieved by eliminating refrigeration, there is the economic bonus of freeing capacity in costly refrigerated vessels for the transport of other more perishable commodities. Although difficult to quantify at this stage of development, the potential financial rewards should be great.

The Wolfson grant for this project was made at the point where exploitation of basic research was just beginning, and at a rather later stage than was the case with the City University gas-sensor development. In both cases, the main factor in assuring success was the recognition by the research groups involved that there were opportunities in a market which had considerable scope for expansion.

The coating materials are edible food additives which have been accepted for use in most major fruit-producing countries, and which appear to delay the development of certain organisms which attack fruit, as well as acting directly on the ripening mechanism. The underlying processes are now under investigation, and the research has been extended to a wide range of tropical and temperate products. The project was given one of the EPIC awards in the first competition run by the Department of Trade and Industry for co-operation between universities and industry.

The capital installation and equipment provided by the Foundation has made post-harvest work a regular part of the teaching and research programme of the Cambridge Department of Applied Biology. One new departure is co-operative research with a consortium of farmers on the storage of crops such as potatoes.

NEW OR IMPROVED PROCESSES

Instruments

Manchester University, Department of Medical Biophysics,
1976, £156 000; 1982, £140 000: Image analysis

The Wolfson Image-Analysis Unit at Manchester has developed the so-called Magiscan system of microscopy in order to automate the scanning of micrographs. A number of successful results have been achieved, some with applications in medicine and others in industry. It is the sole example in Table 3.1 of a project whose purpose is to divise an instrument based upon a new process.

The main medical interest has been in the development of a software package for the automated analysis of cine-radiograms of the heart. The overall effect has been to improve the signal-to-noise ratio, and make possible the storage of sequential frames to simplify subsequent analysis. Automated analysis for the display of chromosome abnormalities has also been developed, and is being extended in a Magiscan 2 package now being brought into use in a well-known European hospital. This work has recently led to a major development contract. The technique is also applicable to the automation of cervical cytology, where the software package is used to pre-screen cervical smears to expedite the selection of those slides which reveal some abnormality, and which call for more detailed attention. The development of the software in this latter instance has proved troublesome.

Two examples of industrial interest in the Magiscan system are the automatic counting of dangerous asbestos fibres in air samples, and secondly the automatic extraction of numerical information from maps, plans, and engineering drawings, for subsequent computer storage and retrieval as and when necessary. The asbestos-counting project has attracted support from industry, and has been completed successfully. The second and more complex project is proceeding more slowly. A further potentially valuable application is in the examination of sub-assemblies in the manufacture of motor cars. Brake-drum assemblies are being considered initially, but image analysis looks as though it will be useful in many inspection operations. Recently close links have been established with a company that will manufacture systems specifically designed by the Unit.

Since the processes that are being developed by the Image-Analysis Unit are widely applicable in medicine and industry, they are being marketed by a University company which can now support itself on licence fees and income from sales.

Mechanical engineering

Two technological projects that have been supported at Aston and Birmingham Universities, and whose histories are comparable, are the use of ultrasonics in the deep-drawing of metals, and the development of forging devices.

Aston University, Department of Mechanical Engineering,
1976, £100 000: Ultrasonics in metal forming

The Wolfson grant to Aston was intended to help the University with the industrial exploitation of the results of some experimental work that had been successfully carried out in the laboratory. The group concerned put up a project for the design of ultrasonic devices for industrial purposes in the first place, but with an eye to more general uses as well. They were confident that the benefits which they had achieved on a small scale could be achieved on the shop floor. They were right.

Metal tubes can be manufactured by rolling, welding, or, as in the Aston studies, by drawing a hollow metal shape through a constricted annular orifice between a die and a plug. Starting in 1971, the Aston team had shown that in the drawing process, the forces needed to extend the tube as it passes between the plug and die could be reduced by as much as 45 per cent by the application of a high-frequency oscillatory force at right angles to the plug and die—and thus at right angles to the direction of drawing. But as late as 1977, potential users remained reluctant to make the speculative investment necessary for developing and installing new machines to utilize the process. It is not the usual business of the SERC and the NRDC to help overcome this kind of difficulty, so that up to that date the problem of transferring to industry the knowledge that had been gained at Aston had not been solved. Since the Wolfson Foundation always remains ready to consider applications on their merits, a grant was therefore made to help the stage of exploitation. It allowed for the appointment of a manager to supervise the transfer to industry of the technology the University was developing.

Both the manager and the senior academic, who had originated the idea,

then made contact with several companies to explain how the use of ultrasonics in metal-forming could improve the processes they were using. The
performance requirements for ultrasonic equipment were then drawn up in
consultation with a company which manufactures a special range of tubes.
It was agreed that a design should be prepared jointly by the University and
the company, with the cost of procurement being shared on an equal basis,
and with the University providing an operating schedule. Difficulties then
arose, particularly with the reliability of the ultrasonic equipment that was
already available on the market when used in the arduous environment of a
tube-forming mill. In practice, the lubricant that is used to assist the drawing process may decompose or degrade to produce an acidic fume, which
may then be drawn into the cooling fans of the ultrasonic generator with
consequent damage to the electronic components. Difficulties of this kind
can of course be rectified by changes in design, and the University has now
produced transducers which are likely to prove satisfactory.

Although further tests of reliability are required, the company with which
the University is collaborating has installed and operated the equipment
which now functions in a normal commercial production manner on a
formal shift system. A paper on the subject was recently published in the
Report of the Metals Society Conference on the Drawing of Metals.

With the ending of the term of Wolfson grant, the members of the project
team have dispersed but, with the modest funds remaining, and in collaboration with several companies, departmental staff are continuing to
make and 'sell' units.

Birmingham University, Department of Mechanical Engineering, 1968, £66 700: Advanced forging machine

'The experience of the Birmingham University team, which was concerned
with forging techniques, has been much the same as that of the Aston group.
This team was given a grant in 1968 to develop an advanced forming
machine, consisting of three Petro-forge hammers linked to each other and
to a furnace by standard automatic-transfer machines. The whole complex
was designed and constructed in accordance with the terms that had been set
out in the application to the Foundation. By the early 1970s the system was
operational. During testing, some parts were found to require re-design,
and funds ran out before the further developments were completed. As
already stated (p. 15), the Wolfson Technological Projects Scheme is not
intended to cater for open-ended research, and once the grant had been
expended, the Birmingham group had to look elsewhere for support. Applications to other agencies for financial help did not succeed and, as a result,
work slowed down and an effective machine as originally conceived did not
materialize.

However, during the 1970s the machine that had been built was used to study the life of dies, and to devise ways of suppressing the noise arising from impact in metal-forming operations. This also had long been part of the Department's research programme. The machine therefore became a research tool, and as such performed a useful and important function. The SERC has provided substantial funds partly for the exploration of the fundamental mechanism which leads to the generation of intense noise on impact in machines such as forging hammers, and partly for the development of computer-aided techniques for the design of hammers that would emit less noise.

It did not take long, however, to decide that although the work on noise suppression that had already been done might well have some 'spin-off', finding a way of reducing noise was going to take too much time and trouble, and that such research would be rather unproductive. The team therefore redirected its work to devise ways of completely automating the forging process, thereby removing the operators from the immediate environment of the machine, and so making environmental noise problems less important. The Department is now pressing ahead with the development of a multi-station forging machine, particular attention being paid to the use of microcomputers and microprocessors to control the forming machinery, and to the use of computer-aided design in the manufacture of the forming dies.

Although the original conception of the forging machine is unchanged, the application of computer technology to its control has revealed new possibilities. The machine has not yet been taken up commercially but, even though costly, it has performed well as a research instrument, and without it several other research projects would never have received financial support from the SERC.

Birmingham University, Department of Mechanical Engineering, 1978, £120 000: Forming techniques

A second two-year Wolfson grant was awarded to the same Birmingham Department to found a consultancy to help transfer to industry knowledge of new technological processes related to the bulk forming of materials. Although the sum that could be awarded for this project was less than was required if the unit were to become self-supporting in a period of five years, the unit was able to spread the award and, by earning substantial fees, to survive. It is now making good progress.

The initial objective was to interest industry in the forming techniques which the Department had developed. Because of the poor state of the forging industry, this aim was soon extended to cover other areas relevant to manufacture, such as automation and computer control, bonding of structures, noise control, robotics, etc. The consultancy unit now services all research and development activities in the Department, and in the five years of its existence, has dealt with hundreds of enquiries, and has negotiated scores of contracts.

Much effort has been devoted to the development of a hardness-assessment computer technique. This has now reached the stage of commercial exploitation. There is considerable industrial and commercial interest in a superplastic moulding process for producing cheaper dies for plastics and thin metal parts, and small contracts have been arranged. The unit has also contributed in a major way to the commercialization of the computer-aided design and manufacture (CAD and CAM) of forming dies. A low-cost robot, designed and developed in the Department, and special gripper designs, have been licensed through BTG. Contracts for the design and development of an advanced robot are now being negotiated. Other projects involve testing, design consultancies, and the negotiation of co-operative research projects involving several companies. The unit also organizes seminars for industry on robotics, bonding of structures, and CAD of forming dies.

Cranfield Institute of Technology,
1980, £52 000: Robot control of warehouse stocks

The work proceeding at the National Materials Handling Centre at Cranfield is listed in Table 3.1 under the heading New Products. In some respects, however, it could be regarded as an example of the class of projects which have as their aim the design of a new process. The objective is to develop a robotic machine for 'picking' (selecting and withdrawing) food packages from warehouse stores. In addition to the mechanical development that is obviously involved, the other aspect of the project concerns the logic of the computer storage system and its relation to the control of the machine.

Salford University Industrial Centre,
1980, £105 000: Warehousing methods

A grant made to Salford is also concerned with warehousing methods and with techniques of 'picking'. The first stage of the project included the creation of a 'Warehouse Classification and Coding System' to assist in warehouse planning, and which could be used to establish criteria for specifying automatic handling techniques required at the picking face.

Current work is leading towards the use of robots for depalletizing and order-picking. This work will culminate in a working model of a warehouse-picking interface, with specific emphasis being placed on end-of-arm tooling and software for merchandise retrieval.

The differences between the Cranfield and Salford projects, neither of which has yet proceeded far enough to enable their prospects to be assessed, are in the approaches to the solution of a problem which is of considerable importance in commerce.

Nottingham University, Department of Production Engineering and
Production Management,
1972, £44 560: Automation devices and control systems

An earlier approach to robotics was encouraged in 1972 at Nottingham. The
Department of Production Engineering and Production Management had
been working in the field since 1964, but up to 1972 the industrial application of mechanized techniques of assembly had been disappointingly slow.
The Department wished to specialize in this work, and it was hoped that the
formation of the Wolfson unit would stimulate industrial interest in the
development of specific automation devices and control systems. The unit
was also designed as an advisory service to industry, and a consortium of
firms was formed to receive and exchange information on developments in
automation. Within the University itself, experimental work continued on
the development of control systems for new types of automatic devices,
leading to significant developments on modularity, programmability, and
versatility for smaller-batch applications.

The unit at Nottingham soon began to resemble a research association
which specialized in particular techniques and which generated enough
income to keep itself going. It came as no surprise, therefore, when in 1980
the Head of the Department was appointed Director General of the Production Engineering Research Association of Great Britain (PERA). PERA is
an independent research organization which is owned by its members. It is
not directly supported by the Government, but is eligible to compete for
Government contracts. Members of the unit were then recruited to form the
nucleus of a small team to operate a three-year Robot Advisory Service on
behalf of the Department of Trade and Industry. This involved setting up a
demonstration centre, running seminars and discussion group meetings,
undertaking feasibility studies, technical and economic appraisals, and
implementing projects for companies.

When the Government contract came to an end in December 1982, the
team concerned formed a consultancy unit which is promoting and undertaking projects for industry concerned with the application of robots and
the introduction of Flexible Manufacturing Systems.

This is an example of a programme, first intended to introduce new
industrial processes, which has moved away from the University and which
has taken the form of an advisory and implementation unit of the type that
is considered in Chapter 4.

Software development

Cambridge University, Department of Mechanical Sciences,
1970, £110 000: Computer graphics

The Wolfson unit at Cambridge (Cambridge Industrial Unit) was founded with a grant that was made independently of the Technological Projects Scheme. It is therefore not listed either in Table 3.1 or in the Appendix. The £110 000 donated by the Foundation in 1970 was initially intended to support a group of staff for five years, or up to the point at which they could become a self-sustaining unit. The most important achievement has been the development of a system of computer graphics, starting from University research, and the licensing of it to Control Data, Delta CAE Ltd., and to Prime Computers CADCAM Ltd., and through them to sub-licensees both in the UK and Germany. The income from this now covers the total running costs of the Unit.

The staff of the Unit have also spent much time in making firms more aware of what University departments in Cambridge can do to help industry.

At 31 July 1983, the capital of the Unit had grown to over £314 000. Fixed assets include a computer, a numerically controlled machine tool, and other hardware, worth about £100 000.

University of Wales, University College, Bangor, School of Engineering Science, 1978, £55 950: Microwave integrated circuits

Another good example of the help that the Wolfson Scheme has provided in the field of software design is a project at Bangor for a specialized study of microwave integrated circuits. The customary sequence in design starts with a performance specification, from an examination of which an outline prototype circuit is sketched out. Analysis indicates whether the circuit will meet the specification, and reveals any deviations from the required performance. The design of the circuit is then modified, and again analysed for performance, the process being repeated until a satisfactory and acceptable response is obtained. Such iterative techniques are familiar but can be time-consuming. The aim of the Bangor project is to evaluate procedures which, with the aid of a computer, can meet specifications more quickly and accurately. The operator may choose either an optimization or a comparison mode, and print-outs of the position reached at any time can be extracted regularly. This considerably speeds up the design of complex circuitry.

Much of the basic work has been completed. Experiments have also been carried out to verify the results, and have proved satisfactory. The larger producers of microwave devices have their own considerable skills in this field, and the College may well have to select a suitable specialist field for application of their work.

The Bangor group has completed a considerable amount of basic work, including practical experiments to verify the results obtained from the programs that have been developed. However it remains to be seen whether the procedures will find wide use. The large producers of electronic equipment have their own ways of doing things.

London University, Queen Mary College, Department of Electrical and Electronic Engineering, 1974, £93 000: Use of paper scrap

The aim of the Wolfson Recycle Unit at Queen Mary College was to introduce new processes into the paper and plastic industries, and in particular to make greater use of scrap materials. Initially linear programming methods were applied to determine how to optimize the input of new raw materials or, alternatively, to recycle scrap in paper manufacture. The paper-making industry is a complex one, and can be subdivided into areas of activity which could be regarded as internal or external to the company. Customers are 'external', since they receive the product. Paper-making factories are also external, since they are places where raw material enters the manufacturing system. Stockholders and dealers are intermediaries between the two external units, and are therefore regarded as 'internal' districts. Materials flow in certain patterns between these various types of unit, and within each unit there are other patterns of flow.

The intention of the work was to optimize these flow patterns within and between districts, and data about these flows, already available within the paper-making industry, were used to design a computer model to indicate the most efficient way of moving materials within the system under different circumstances. The original intention had been to follow this work with a similar study of the plastics industry. This idea was abandoned when it became apparent that the technique could be used to develop an energy and resources model of much wider application.

A link was then established with the European Commission for a large-scale study of prospective energy supply and demand within the EEC up to the year 2020. The model had to be easy to use, yet sufficiently versatile to deal with revised scenarios as and when necessary. The QMC program meets these requirements, and an income from contracts of some £100 000 per annum has already been realised.

The computer model that has emerged from this study has a wide range of applications and should attract sufficient income to maintain the group.

University of Manchester Institute of Science and Technology (UMIST), Department of Textiles, 1978, £44 800: Pattern-processing systems

The object of fabric pattern processing is to transfer a designer's sketch or drawing into coded information which can be held in store and extracted when necessary to control fabric-production machines. At the start, the aim of this project was to develop low-cost pattern-processing systems for use in all sectors of the textile industry. At the start the work focused on knitting, but it has since been directed to weaving, to carpet-making, and to printing processes. Most recently it has been developed as a tool to help designers explore the effect of colour on design.

The system is applicable to most kinds of textile design work, with processing times being reduced by as much as 90 per cent over previous manual methods. The new development is compatible with a range of design-processing systems. Its capital cost is far lower than existing automatic scanning systems, with which it compares favourably in performance and ease of use. The designers do not have to understand computers, since they can devise programs using a 'menu' of textile designers' terms.

Considerable interest was aroused when the Textile Design Conversion System was exhibited at the four-yearly International Textile Machinery Association Fair held in Milan in 1983. A new company has now been set up, supported by the British Technology Group, to market the product, and to assist small textile companies. This will, at least in a small way, help increase employment in the area.

Materials technology

Birmingham University, Department of Minerals Engineering,
1970, £128 200: Metals from scrap

The extraction of valuable metals from secondary sources (scrap) was the purpose of another Birmingham University project which was supported under the Scheme. Its main aim was the recovery of copper and associated metals from low-grade scrap, using existing research in pyrometallurgy, hydrometallurgy, and analytical techniques.

The usual pyrometallurgical method of obtaining copper from mixed scrap is by a two-stage process, in the first of which, carried out in a blast furnace, undesirable metals such as iron and aluminium are oxidized and taken up by slagging. The remaining impure copper and other metals are then reduced in a converter to give a suitable material for electrolytic refining to pure copper. The Birmingham work was intended to replace these batch processes by a more economical continuous process in which metal and slag move in counter-current. Impure metal enters the furnace at an oxidation zone where iron and aluminium are removed, as in the blast furnace, by slag. The remaining impure copper then proceeds along the furnace to a reduction zone where a sufficiently pure copper for refining is produced as in the converter stage of the batch processing.

The new Birmingham scheme is largely self-sufficient for energy needs, continuous in operation, and amenable to automation. It should replace the conventional batch process of oxidative blast furnace followed by a copper converter. The latter recovers copper suitable for electrorefining, but other non-ferrous metals are lost. The next step should be the scale-up of the furnace to 0.5 tonne/hour capacity, but this is unlikely to be done until general economic conditions improve.

Hydrometallurgical methods were used to separate the metals in the alloy which the furnace produced. Extensive studies showed that acid chloride media were most convenient and effective for dissolving the metal. Further research was carried out into the behaviour of cation-exchange extraction reagents in contact with chloride media. A scheme was then developed for the selective recovery of the individual non-ferrous metals by solvent extraction, using existing extraction reagents dissolved in kerosene. In this way copper, nickel, zinc, lead, and tin have been recovered, the last metal often contributing a value equal to that of the major component, copper.

Liverpool University, Department of Metallurgy, 1974, £61 840: The use of plastic scrap

Recovery of useful material from waste was also the purpose of a grant given to the Department of Metallurgy at Liverpool, but in this case the raw material was plastic scrap. A source from which waste material can be collected cheaply and in large quantities must be assured if there is to be any likelihood of commercial success in upgrading plastic scrap. Moreover, since the cost of plastics is sensitive to that of oil, the economics of recovery from lower-grade wastes will also be dependent on the price of oil. One significant source of plastic waste material is the cable-making industry, where it has long been the practice to recover metal from scrap cable.

The project team in the University was fortunate in so far as it was able to make a collaborative arrangement with a suitably large cable-making company situated close to Liverpool. The company already possessed excellent facilities for reclaiming waste metal. The plant it had installed has now been extended to allow for the reclamation of plastic insulating material. The contribution of the company to the project has been important. It used its own engineering experience to build a pilot plant, which provided data which could then be used to design a largely automated full-scale unit to treat the scrap it produced. This new plant has operated on a weekly basis with very little trouble for over two years, separating polyethylene, PVC, and any metal which may appear in the scrap. The process has now been marketed by the cable-making company. The patents have produced royalty payments for the University, which has returned part of its grant to the Foundation for the support of further technological projects. The company estimates that the recovery of the copper alone justifies the cost of the plant, quite apart from the value of the 25 tonnes of PVC granules which it recovers every week.

Liverpool University, Department of Metallurgy, 1976, £120 000; Birmingham University, Department of Metallurgy and Materials, 1983, £100 000: Plasma nitriding

In 1976 a Wolfson Plasma Processing Unit (WPPU) was established at

Liverpool University. Five years later, in 1981, the leader of the Unit was given an appointment in Birmingham University, and took the Unit with him.

The primary aim of the WPPU, as envisaged in 1976, was to provide a vehicle for the direct introduction of plasma nitriding experience into UK engineering industries. In the older nitriding processes, a suitable steel is heated in a current of ammonia. Nitrogen produced from the ammonia combines with the surface layers and gradually diffuses into the metal. The process is time-consuming and may produce distortion of the work piece.

By contrast, in plasma nitriding, nitrogen ions are produced in an electrical discharge and driven directly into the surface layers of the metal by the applied voltage. The process of nitriding is speeded up and the work piece remains cool and free from distortion.

Over the past seven years the WPPU has collaborated with many manufacturing companies, some of which have installed their own equipment. One company working closely with the university has installed equipment costing £400 000 to provide a service for those companies not wishing to instal their own.

With the downturn in the economy, the introduction of plasma nitriding by UK industry has slowed down. The efforts of the Unit for the past two years have accordingly been concentrated on plasma-processing research, especially in the field of plasma carburizing, nitrocarburizing, and boriding.

The Unit has now been incorporated into the Wolfson Institute for Surface Engineering (WISE), which was founded in 1983 at Birmingham University as a separate technological project. The principle aim of WISE is to serve as a centre in the UK for up-to-date information about Surface Engineering, such as chemical vapour deposition, physical vapour deposition, laser surface modification techniques, as well as the original plasma thermochemical treatments. Many of the novel surface-heat treatments and coatings for metallic materials are already of commercial interest.

Salford University, Department of Electrical Engineering, 1976, £96 000: Ion-plating

The ion-plating unit that was formed at Salford with the help of a grant given in 1976 utilizes vacuum plant similar to that employed in the Department of Metallurgy at Liverpool. The unit now provides a service to industry for ion plating with different deposits on a variety of materials.

The unit's electrical discharge equipment is used by the Department for research, and is also available to the University's Industrial Centre for work on behalf of industry.

Apart from processing components to customers' requirements, the unit also undertakes the design of equipment for companies which wish to introduce the process into their own factories. Links with possible manufacturers of the required equipment have been built up as part of the service.

Income has grown steadily since the Wolfson grant ended. The unit is now self-supporting, even though its growth has been slower than was originally predicted.

Energy

Queen's University, Belfast, Department of Civil Engineering, 1972, £54 000; 1981, £80 000: Box-girders and wave power

Of the project grants in the field of energy, that given to Belfast is of particular interest. It is listed in the Appendix as a civil engineering project (construction) rather than as one having to do with energy. This is because the stated purpose for which the original grant was made was for work on composite materials and composite constructions. Research on box-girders showed that their resistance to buckling was greatly increased by filling them with plastic foam bonded to the box-girders. This work was expected to lead to widespread exploitation in the construction industry, but the rapid escalation in oil prices after the Yom Kippur war in the Middle East increased the price of oil-based plastic, of which enormous quantities would have been needed for the filling of box-girders in large structures. These composite constructions are clearly a valuable development, but at the same time too expensive at present for extensive use.

After consultation with the Foundation, the remainder of the grant was therefore redeployed on a programme of research into wave-power as an alternative source of energy. The preferred system was a device with an oscillating column of water, which produced a cyclical flow of air through a novel form of self-rectifying turbine—an idea which has now been patented. A 4.5-metre diameter version of the water column was built and installed for test in Strangford Lough in late 1977. Unfortunately it was damaged in a severe storm before useful records could be obtained. In October 1981 the Foundation made a further grant to expand the programme of work. This grant, and the residue of £30 000 from the original award, should make it possible to develop an effective oscillating water column and the associated air turbine. Since mid-1978 the research team has also received research grants from other bodies, amounting to some £400 000, to help continue basic studies to the point where small-scale applications become feasible.

University of Manchester Institute of Science and Technology, Department of Polymer and Fibre Science, 1980, £113 000: Production and processing of polymers

This project is classified in Table 3.2 under the heading of Energy, since the aim of the work was to conserve energy in the production and processing of polymers, particularly in the expanding polyurethane industry. The programme consisted of two main parts: (i) the development of new non-oil based chemical precursors for polyurethane formation, and (ii) the development of small-scale processing equipment for in-situ formation and moulding of polyurethanes by reaction-injection moulding.

Important results have been obtained. New di-isocyanates have been made from renewable agricultural sources by a process that has been developed and patented by the team. The polyurethanes derived from these sources have properties comparable with those based on oil products.

Polyurethanes are normally formed by the reactions between di-isocyanates and polyols, and additional patents are under consideration on the production of polyols from waste materials and from such carbohydrates as lignocellulose.

Parallel with this work, a machine for reaction-injection moulding has been built within the Institute and patented. It can mix the polyols and di-isocyanates, with or without fillers, to give reinforced and non-reinforced polyurethanes.

Useful links have been forged with various firms, and some industrial funding has been received for further work. Additional staff have been appointed. The unit is helped by the UMIST Research and Consulting Services with patent and exploitation arrangements.

Biochemistry

Fermentation and antibiotics

In 1959, before the regular scheme of technological grants was started, the Wolfson Foundation made a grant of £200 000, to which further smaller grants were added, to the Biochemistry Department of Imperial College, London for the establishment of a fermentation and extraction pilot plant in order to extend the laboratory-scale work on antibiotics of the late Sir Ernst Chain, FRS, and to demonstrate its suitability for industrial use. The plant has been used from time to time to carry out pilot-scale work for industrial concerns, and in this sense has provided a service to industry. It is still in operation, playing an important part in both teaching and research. Plans have also been made for the plant to be made available to an industrial company concerned with the development of drugs.

London University, Imperial College, Department of Biochemistry, 1976, £46 400: Ethanol from sugar

The purpose of a grant made to Imperial College in 1976, under the Technological Projects Scheme, was to help the Biochemistry Department develop new processes of industrial fermentation, in particular for the rapid production of ethanol from sugars. This ambitious project had as its aim the production of a thermophilic micro-organism capable of fermenting sugars at temperatures of 70 °C and higher. By means of genetic manipulation, an organism was produced which grew rapidly at 70 °C, but made little ethanol, giving yields equivalent to those of yeast. At this high temperature, mild vacuum gives 25 per cent (w/v) ethanol vapour and only 4 per cent (w/v) in the liquid phase. Further improvements are envisaged, and the industrial relations that have been forged assure the viability of the unit.

Manchester University, Department of Chemistry, 1974, £68 300: Alternative sources of food protein

It was widely expected in the early 1970s that there would be a shortage of food protein for both human and animal consumption. To help in the search for alternative sources of supply, a number of grants were made under the Wolfson Scheme. One at Liverpool (1974) was concerned with the cultivation of scallops; another at Reading (1974A and 1974B) with the extraction of protein from vegetable sources; while a project team at Belfast (1978) studied the possibility that molluscs might be used as cattle food. Reference has already been made to the Ulster (1978) project on the extraction of protein from whisky residues, and to the project at Bath (1978) for extracting protein from bone. The latter two have proceeded to the pilot-plant stage, and show signs of proving economically viable. The others, although technically successful, are not likely to prove commercially attractive since the expected shortages did not materialize.

Like the Ulster project, a Biomass Unit which was set up in Manchester in 1974, concerned itself with the production of high-protein feedstuff for livestock. In this case the source material was waste from food-manufacturing processes. Although such wastes have the disadvantage of being non-homogeneous, they are often an easier starting point than whisky residues. In the event, the Manchester project, like some others of a similar kind that were supported, faltered because the fuel costs of the process increased to the point where it was cheaper to turn to such alternative primary protein products as soya bean and fish meal.

Another extraneous factor which had an adverse effect on efforts to exploit alterative sources of protein was the relaxation of government controls on the disposal of factory effluents. The energies of the Manchester team were then redirected to new problems, particularly to more efficient fermentation processes for the production of ethanol. Although the work was less innovative than the Imperial College project (p. 45), it none the less attracted funds after the expiry of the Wolfson grant from industry and government, rising to a peak of £70 000 per annum in 1980. The commercial company that had collaborated with the Wolfson Unit produced designs for a pilot plant, but as economic conditions deteriorated it had to withdraw its support.

Recently links have been established with a Canadian University which may result in the development of the ethanol process for the conversion of agricultural waste material, the present disposal of which is causing severe pollution problems.

The future of the Manchester Biomass Unit is now uncertain. If new processes are to be developed, a fresh source of funds will have to be found. The experience of this project is a salutary reminder that although a project may be started with every expectation of filling a real market need, a change in general economic circumstances may well cause the market to contract or disappear.

Sheffield University, Department of Biochemistry and Department of Chemical Engineering and Fuel Technology, 1978, £120 000: Plant-cell technology

There are a variety of biotechnological processes which depend on fermentation. One approach, particularly applicable to the production of some drugs, lies in plant-cell culture. The Wolfson Unit of Plant-Cell Biotechnology in the Department of Biochemistry at Sheffield was set up with two directors, one from the Department of Biochemistry, and the other from the Department of Chemical Engineering and Fuel Technology. The Unit is thus able to carry out a range of work from basic biological research to engineering design.

Plant-cell biotechnology is a field in which there is likely to be much competition—academic, technical, and commercial—and it is important that a strong position should be established in the UK. Consequently, the Foundation decided in 1981 to strengthen the resources for research and production by making a further grant to the Sheffield Unit, but not as part of the Technological Scheme, in order to expedite its work. The University then established a Wolfson Institute of Biotechnology as a separate academic department, with its own Director and staff. Subsequently the interest taken by the UGC in biotechnology led to Sheffield receiving a £50 000 recurrent grant to establish further academic posts in the Wolfson Institute.

The main research programmes are concerned with the utilization of feedstocks by plant cultures; the synthesis of natural products; the technology and biochemistry of mass cultivation of plant cells, including vessel design; and embryogenesis and plant-cell propagation. At present the techniques are based on the use of bioreactors of between 5- and 100-litre capacity, and the Institute's culture bank contains cell lines from about twenty species of plant. Among these are the opium poppy, a source of codeine and morphine; digitalis lanata, yielding the heart drug digitoxin; and cinchona, for the production of quinine.

The concentrations of the products that are being produced are encouraging, and scaling the process up to pilot-plant operation with reactors of 500-to 100-litre capacity is proceeding. The great potential of this new method of producing drugs, which have traditionally been derived from plant sources and their variants, has been acknowledged by the support now being given to the Sheffield unit by industry. The unit has also received substantial contracts for work on biopolymers and for the cultivation of tobacco plant cells. Other objectives include food flavours, dyestuffs, and pesticides for protecting crops.

The need to achieve competitive costs in the production of plant-source materials will require vision on the part of the research workers, as well as ingenuity from their engineering colleagues.

In late 1982 the University was helped by the 'Investors in Industry Group' (formerly Finance for Industry) to establish a new research and development company named Plant Science Ltd. The senior members of the Wolfson Institute became substantial shareholders in the company, the formation of which relieved them of the responsibility for fund-raising, thus permitting them to concentrate on basic and contract research.

More recently, subsidiaries of the main company have been formed in the United States to help exploit the market there. A UK subsidary company, Plant Propagation Ltd., is also in an advanced stage of planning. This company will be concerned with plantlet production through micropropagation for horticulture and agriculture. A site has been acquired not far from the University, and staff are being recruited. It is hoped that this company will become operational as a manufacturing base early in 1984.

4

SERVICES FOR INDUSTRY

All the projects listed in Table 3.2 had as their general aim the provision of services to industry. The representative sample discussed in this chapter comprises one group which involves specific research and development; and a second, which provides sophisticated testing services either within or outside a Wolfson unit or institute.

RESEARCH AND DEVELOPMENT

London University, Imperial College, Department of Mining and Minerals Technology, 1974, £60 000: Surface-active clays for foundries

In 1974, as already noted, preference was given to projects which related to the conservation of energy and materials, and which if successful could lead to some reduction in imports.

One grant was made to an Imperial College team which proposed to find a suitable domestic igneous rock from which to derive smectites, the surface-active clay materials which are used in foundries and steelworks, and supplies of which, such as Bentonite, are largely imported. After a suitable domestic rock had been identified, the conditions required for decomposition were evaluated in the laboratory, first on a few grams of material and later on larger samples. The resulting product proved to be suitable as a binder in foundry sands (one of the main uses of bentonite) and for use in the sintering of iron-ore pellets. It was recognized from the outset that the economic aspects of the project were speculative, and this has proved to be the case. None the less a useful research and teaching centre has been established for work on active clay materials, and one which could provide consulting services to industry.

London University, Imperial College, Department of Metallurgy and Materials Science,
1972, £72 609; 1974, £106 000: Smelting, refining, and recovery

The grant of £72 609 made in 1972 was for research and development on the extraction of metals. The work was concerned with the use of finely dispersed phases in smelting and refining to increase the kinetics of the processes. This has important implications for the designer of process plant. More intensive processes can be designed which are more efficient in their use of energy and capital.

Co-operative projects with industry were successfully carried out in the fields of copper refining, the reduction of iron oxide, the smelting of stainless steels, and the reduction of tin from slags.

Information obtained from such work is not generally patentable, and it is impossible to quantify its value to industry. The return to the research group itself depends on the judgement of industrial collaborators, as reflected in the funds they are willing to offer for further work, and the extent to which bodies such as the SERC value the work as a general contribution to technological knowledge. Since the 1972 Wolfson grant, the unit has obtained more than £600 000 (at current prices) from other sources to maintain and extend its work.

The grant that was made in 1974 to a different Imperial College team was for a project concerned with the recovery of metals from waste materials. For example, vanadium compounds occur in oil-fired boiler ash, alumina in coal and power-station waste, and silver in secondary smelting operations. Interesting work was done on the separation of zinc and lead from iron using a molten two-phase chloride-oxide system. Several papers and reviews were published and a patent application submitted for a new method for obtaining high-purity chromic oxide from chrome ore. Although the recent downturn in the economy has diminished interest in the recovery of metals from secondary sources, the team now provides a consultative service.

London University, Imperial College, Department of Metallurgy and Materials Science,
1978, £125 000: Ceramic electrolytes and solid-state batteries
The Wolfson Unit for Solid-State Ionics was set up at Imperial College to extend the work of a well-established research group which for long had been concerned both with basic research and with the application of knowledge already acquired to developments of direct interest to industry. Two of the latter may turn out to be of considerable importance. The first was the evaluation of oxide-ceramic electrolytes and the establishment, in conjunction with a small firm, of facilities for their manufacture. The microstructural features of, and a manufacturing process for, an electrolyte based on zirconia have been optimized. Samples have been incorporated into oxygen monitors for the control of flue gases in order to improve combustion efficiency.

Collaboration between the Unit, the manufacturer of the ceramic, and an instrument maker, resulted in an improved design of an in-situ flue-gas monitor which is now ready for marketing. This collaboration has also resulted in the establishment within the UK of a manufacturing facility for zirconia-based ceramics. With the help of DOI and EEC contracts, this is now being expanded to provide toughened ceramics for adiabatic diesel engines and to manufacture wear-resistant materials.

The flue-gas monitor that has been designed could be compared with the metal-air cell that was produced by the City University unit and referred to on page 28. The two devices will, however, serve different requirements and operating conditions. Unlike the City University team, that at Imperial College has been more concerned with research and development than with actual manufacture.

The second project undertaken by the Wolfson Unit for Solid-State Ionics, and one which may prove to be even more rewarding financially, is the application of new ionic solids to the design of solid-state batteries. These are of world-wide interest, and if British industry is to play a part in their development, it is important that there should be an underpinning of research by the universities. The Wolfson grant has provided cash for equipment and, together with funds for staff provided from other sources, has helped establish a research and development team which in due course can move on to other related industrial problems.

The group continues to thrive. It usually has a membership of about 15 research staff supported by funds from a variety of sources. In the past five years some forty papers in the field of solid-state ionics have been published. The Unit is now recognized as one of the principal centres in the world for research and development activities on new material for energy conversion and conservation. It also acts as consultant to the EEC.

Newcastle University, Sub-department of Crystallography,
1970, £155 850; 1981, £100 000: Engineering ceramics—'Sialons'

In 1970 a grant was made to Newcastle to establish a research and development group for the characterization and production of new high-strength metallic and non-metallic materials. This award is classified in Table 3.2 under research and development, but developments since the end of the initial Wolfson grant could well have justified the inclusion of the project in Table 3.1 under New Products.

A Sub-department of Crystallography had been set up in 1964 as part of the University's Department of Metallurgy. In 1968 it was predicted that silicon nitride, then emerging as a material with useful high-temperature engineering properties, might be the forerunner of a whole family of potentially useful ceramic compounds that incorporated nitrogen. The Wolfson grant of 1970 gave an additional impetus to the programme of work on engineering ceramics, which combine strength at high temperature with resistance to oxidation.

Many such new compounds, now known by the acronym 'Sialons' have since been derived from silicon nitride and silicon oxy-nitride by the simultaneous replacement of both silicon and nitrogen by aluminium and oxygen. The family of compounds continues to grow, with silicon carbide or aluminium nitride taking the place of silicon nitride. The work has aroused world-wide interest and, although much basic research is still needed, the Wolfson group is already collaborating with an industrial firm.

The first successful commercial application has been tips for cutting-tools, where the oxidation resistance and the mechanical resistance of 'Sialons' to wear and shock can be markedly superior to that of tungsten carbide. Resistance to oxidation is good up to 1400 °C. Similar techniques can be used to make dies for tube- and wire-drawing. The advantages over established materials are clear, but there remain problems of manufacture and cost.

Apart from the work which it has done on its own, the firm which is col-laborating with the University has contributed £150 000 to the Sub-depart-ment of Crystallography. About £250 000 has also been obtained from other sources, mainly the SERC, to further the Sub-department's researches. In addition the Wolfson Foundation made a further grant in 1981 to accel-erate the work. In the same year it provided a grant to a team in Warwick University whose work in the same field also promised to find an industrial application.

University College, Cardiff, Department of Electrical and Electronic Engineering, 1968, £132 000: Soft magnetic materials

One of the earliest of grants in materials technology was made to the University of Wales' Institute of Science and Technology (UWIST) at Car-diff for a study of soft magnetic materials. From the start, the main aim was to find industrial uses for new materials, and in the early stages a particular interest had been shown in applications to large power transformers.

The six scientists who started the Institute later moved to University Col-lege, Cardiff, where they enlarged their programme of work. Studies of transformer core materials continue. New materials such as amorphous magnetic ribbon are being investigated. The magneto-resistivity properties of thin-film ferromagnetic materials have been used in the development of such devices as magnetic card-readers, security devices, and metal detection systems. Some are at the point of exploitation by local companies.

The strength of the project team is impressive. The quality of the basic research it carries out is high, and its scope has not only grown with the aid of the Wolfson grant, but also with the financial support provided by many other sources, including SERC and British firms. Since 1970 the group has published more then 170 noteworthy papers.

Sussex University, Department of Electrical and Electronic Engineering, 1972, £127 655;
Warwick University, Department of Electrical Engineering,
1972, £147 750: Magnetic levitation of vehicles

These two complementary projects on the magnetic levitation of transport vehicles are pathfinding studies for a major engineering enterprise. Both were started when it was impossible to make any confident forecast of a British market for the ultimate product. British Rail showed an interest, but without indicating that any immediate commercial application was possible. BR has also carried out work on the subject in its own laboratories.

The Warwick group was primarily interested in the development and exploitation of superconducting magnets, which can now be produced for a wide variety of applications. Research at Sussex focused on the use of controlled direct-current electromagnets, and the history of its enquiries exemplifies the problems which can affect all workers in this field. The stiffness of an electromagnetic bearing depends both on the strength of the magnetic field and its shape in relation to the duty to be performed. To ensure stability, any departure from the optimum position has to be detected by sensors, and suitable adjustments have to be made to the bearing to return it to the ideal position. Speed of response, damping, and other factors are obviously important. There are therefore a number of problems of control in addition to those which arise in the more conventional forms of rail transport.

The driving force for a magnetically suspended vehicle would probably be a linear-induction motor, and a small vehicle of one tonne in weight, using such a drive, was constructed and installed in the laboratory at Sussex. Smaller test carriages for proving detailed designs based on the suspension concept have also been built, and similar small units have been developed at Warwick to run on a 600-metre test track.

Studies of magnetic levitation have passed through many of the essential first stages of research. Practical development could follow if suitable applications can be found. Some encouragement has been given by the proposal to install a system of this kind at the National Exhibitions Centre in Birmingham.

The Wolfson grants came to an end in 1976, but both Universities have maintained their interest in magnetic suspension and related magnetic devices. Several other schools have investigated linear motors and other power sources. The Sussex group has been kept going with the help of university funds and SERC grants. It has also received support from industry, and has made money from the sale of magnetic devices. In 1981 its work led to a successful application in the design of electromagnetic bearings for machine tools. In collaboration with a manufacturing firm, the Sussex group has also produced the prototype of a heavy industrial centrifuge using magnetic bearings and a linear induction motor. The tubular rotor, which is suspended by magnetic bearings, has a diameter of 1.6 m and weighs 8 tonnes. A comparable machine without the magnetic suspension would rest on roller bearings and, being driven mechanically, would be very noisy. Its power consumption would also be four times that of the magnetically suspended machine. The new design is now undergoing industrial trials.

Many other uses for magnetic bearings are being considered. When applied to machine tools, the stiffness of the bearing could be controlled to match the requirements of the load. The controls would not differ markedly from those developed for the heavy centrifuge, nor from those required in the original concept of a magnetically levitated train.

Arising out of its work on magnetic suspension the Sussex team has moved into the field of instrumentation systems.

WOLFSON UNITS, ADVISORY AND TESTING SERVICES

Southampton University

Southampton University is unique in being the home of twelve industrial units, eleven of which were formed with the help of Wolfson grants awarded between 1968 and 1981, and amounting in all to £970 000. This includes a £250 000 contribution (outside the Technological Projects Scheme) towards purpose-built accommodation for some of the units.

Between them the nine Wolfson units employed sixty staff in the University financial year 1982/83 and earned £1 189 000. One of the first, the Centre for Noise and Vibration Control, which was established in 1968, had an income that year of £284 500. The Auditory Communication and Hearing Conservation Unit, and the Cryogenics Unit, which were formed only in 1981, and are therefore not shown in Table 3.1 or 3.2, earned £140 000 and £94 000 respectively.

In general all the Southampton units have the same administrative structure. They were set up when there were clear opportunities for marketing the results of research in progress in the University Departments concerned.

The individual units are not as strongly linked together as they are with the University Departments from which they developed. It is a pattern which has produced a valuable and extensive range of services for industry, and one which could readily be followed in other universities.

Southampton University, Institute for Sound and Vibration Research, 1968, £30 000: Noise and vibration control

The longest established of the Southampton Wolfson Units is that for Noise and Vibration Control. When it was set up, work in this field was being carried out in the existing Institute for Sound and Vibration Research, which had already achieved a national reputation. It had also established an industrial noise unit which carried out contract work to meet industrial needs.

The purpose of the Wolfson grant was to provide initial support for a professional group which would concern itself solely with industrial problems. Staff with appropriate experience were recruited, and a small laboratory was established both to service and maintain equipment for experimental work and to provide facilities for analysing data. The unit has been fortunate in that during the fifteen years of its existence the problem of noise in factories and in the environment at large has become a matter of increasing public and governmental concern.

On the other hand, the demand for consultancy services has declined in recent years as a result of the economic difficulties with which so many firms have been contending. As a consequence the relations between the unit and academic staff have suffered because the unit has had to devote so much of its time to increasing earnings in order to stay in business. Any weakening of unit/industry links diminishes the purpose of housing a unit within the environment of a university. The Noise and Vibration Control Unit has managed to survive the difficulties of the past few years mainly because of its close links with the academic staff of the related Department, who have supported its industrial work. The Unit can be regarded as the industrial arm of Southampton's Institute for Sound and Vibration Research which, although dealing with applied science and engineering, is mainly concerned with longer-term research. Such a relationship between Wolfson units and the academic groups from which they sprang is always a source of strength.

Southampton University, Departments of Aeronautics and Ship Science, 1970, £16 500: Marine technology and industrial aerodynamics
Like the unit for Noise and Vibration Control, the Wolfson Unit for Marine Technology and Industrial Aerodynamics has its roots in work that was started in 1968 in one of the University's departments. In 1970 the Foundation provided a grant to develop the Department's work by setting up a unit whose original objective was to develop a technical support service for firms making small craft. The grant provided for specialized services, including computing and help in the structural design of plastic boats. These are services which small manufacturers need to purchase from outside their own company, a need which a university can often provide. As with other units, the industrial recession has demanded considerable reorientation and broadening of the Unit's fields of interest in order to ensure survival, and this has been successfully accomplished.

The Unit of Marine Technology retains a foot in, and has widened the scope of research in, both the University's Department of Aeronautics and the now-separated Department of Ship Science. It also has close relations with a wide range of domestic as well as some overseas customers (25 per cent of its work is 'export'), and it has earned enough not only to assure a future for the staff, but also to allow of the accumulation of reserves. Its own work programme covers a wide range of industries, from offshore oil to the aerodynamic design of racing cars. Its achievements were marked by the award of a medal to the Director of the unit by the Royal Institution of

Naval Architects. The Unit's income in 1983 was £277 000, a 60 per cent improvement on the previous year.

Southampton University, Department of Electrical Engineering,
1970 £24 800: Applied electrostatics

The Wolfson Applied Electrostatics Advisory Unit was formed in 1971 with the help of a grant which was made in order to assist the Department's Applied Electrostatics Research Group in their consultancy work. This group had long been engaged in contract research, but had found it difficult to carry an additional and variable load of short-term consultancy enquiries, which took anything from days to months to answer. As there are few university research teams concerned with applied electrostatics, and even fewer industrial liaison units, work has come to the Southampton Unit without the need for much marketing.

The Unit, which began with a staff of one and which now numbers eight, undertakes work relating to the applications and the hazards of electrostatics. Enquiries about applications concern such topics as powder coating, crop spraying, precipitation, fuel atomization, and printing. Projects relating to hazards include pumping of fuel and inflammable liquids, minimum ignition energy of powders, pneumatic conveying, and storage systems. Of particular note is a unique silo installation which enables assessments to be made of the electrostatic hazards that can arise in the transport of powders. Approximately £250 000 has been invested in this project by fourteen commercial sponsors (at home and abroad). The Health and Safety Executive has also contributed to the costs. Recently the Unit has expanded to provide a service in electrostatic instruments. Its income in 1982/83 was £134 000.

Southampton University, Department of Aeronautical Engineering,
1970, £50 000: Engineering materials

The Engineering Materials Consultancy Service at Southampton has developed in a slightly different way. It was given a Wolfson grant in 1970 to provide a special service for the selection, testing, and use of materials in engineering applications, and for the investigation of cases where the unsatisfactory performance of equipment, components or products was attributable to the characteristics of the materials of which they were made. The unit is provided with first-class instruments and machines, of a type not usually available to small and medium-sized industrial concerns outside the metallurgical and material industries. The annual turnover has fluctuated from year to year; income in 1982-83 topped £41 000.

Southampton University, Departments of Organic Chemistry and Biology,
1976, £136 000: Biochemical pesticides

In 1976 a Wolfson grant was made for a joint proposal from the Departments of Organic Chemistry and Biology to establish a Unit for Chemical Entomology. The primary aim of the project was to identify and synthesize natural compounds which are either toxic to insects or which serve as attractants or repellants. A self-supporting advisory service was also to be organized. Unfortunately, the attempt to identify commercially profitable areas of work and to secure overseas contracts was at first unsuccessful. The interest of some large companies was aroused, but little financial support was forthcoming.

A contract to develop a domestic fly-trap was then obtained from a company which was too small to maintain an effective research effort of its own. This project was partially successful. Subsequently a major collaborative effort was initiated to assist a small company in its use of synthetic pheromones for controlling agricultural pests, including fruit pests. As a result of this collaboration, the unit is now selling control systems in Europe. A cockroach trap that was patented worldwide, and first marketed in 1981, owes features of its design, including a specially developed attractant, to the unit. Other studies have included control of the olive fly, the pine beauty moth, and other forestry pests which are exceptionally difficult to eradicate. Although the total income of the unit has fallen short of expectation, the knowledge and experience gained in the course of its work has led to a substantial number of research publications, and established an international reputation for chemical entomology in Southampton.

Southampton University, Department of Chemistry,
1978, £162 660: Electrochemistry

Faster progress has been made in the Wolfson Centre for Electrochemical Science, which was established as a consultancy service to give short-term help to companies, and to seek contracts for longer-term research from companies with insufficient resources to do their own. Subjects already worked on include the electrochemistry of molten salts; the use of electron-chromatic compounds for electronic display applications; engineering problems such as the design of reactors for electrochemical processes; the development of a trickle-tower reactor for use in metal reclamation; the removal of pollutants from effluent liquors; and battery research and corrosion problems. Work on the trickle-tower reactor was helped by grants from the British Technology Group, and four different reactors have been developed to the small pilot scale for several firms. A satisfactory balance between research and industrial development has been established by the Centre, which in 1982 earned more than the size of the grant by which it was set up. Income seems likely to grow sufficiently to maintain the existing staff of six research scientists.

Southampton University, Department of Electronics,
1968, £24 400; 1976, £25 500: Microprocessors in manufacturing processes
Southampton's Electronics Department has received two Wolfson grants,
the first in 1968 to set up an industrial liaison unit, and the second in 1976,
more specifically to encourage the use of microprocessors in manufacturing
processes. At the start a number of small projects were undertaken, some in
the form of consultancies, some on the testing of components, and others
which involved the design and construction of special equipment. An example of the latter was the application of microprocessors in the design of an
electronic control system for a grinding machine used in the cutting of glass
vases and wine glasses. To date six machines with this novel form of control
have been successfully installed, and of these two are in Japan and two in
Sweden.

During the five years which followed the 1976 grant of £25 500, the unit
earned £350 000. It was, however, relying on one or two large contracts and
this, combined with the onset of the recession did cause problems. Because
of the recession, it was decided to accept only contracts which related to an
active interest of the Department of Electronics. Recently work on optical-
fibre communications has provided a substantial proportion of the unit's
income which, in the year 1982/83, amounted to £61 000.

A further change was the formation in 1981 of a Microelectronics Unit
whose work focused on microchip design. This Unit earned £18 000 in the
year 1982/83.

Edinburgh University, Department of Electrical Engineering,
1968, £130 700: Microelectronics consultancy service
Several university electronics departments other than those at Southampton
have also been supported by the Foundation. One of the first grants given
under the scheme was to the Department of Electrical Engineering in Edinburgh, to help it establish a unit to assist local industry. The project has
been outstandingly successful. The Department already had a considerable
reputation for fruitful research; microelectronics was the basis of a com-
paratively new and growing industry; and Scotland was not only attracting
new enterprises but was already the home of a number of expanding electronics companies. As well as seeking wider contacts, a new unit was
therefore able to provide a consultancy service to meet the needs of several
potential customers. Because of the breadth of electronic expertise required,
the backing of a vigorous academic department was essential to the success
of the unit.

The considerable help which the Wolfson Microelectronics Institute at
Edinburgh has given to industry in the Edinburgh-Glasgow region was
recognized by the Lothian Regional Council in 1979, when it agreed to create
a chair in Microelectronics in the University, and to set up and equip a special

microprocessor laboratory to help local firms develop new products. It was agreed that the teaching potential of the Wolfson Institute should also be exploited by way of the secondment of a senior science specialist from the Lothian Education Department to establish links with schools in the area.

The original grant from the Foundation in 1968 amounted to £130 700. The Institute now employs twenty people, and its income is about £500 000 a year. Since its foundation, total contract income has amounted to some £2 million with, it is to be expected, commensurate financial benefits to the firms that have used the Institute's services. Further information about the Institute is given in the next chapter.

Essex University, Department of Electrical Engineering Science, 1972, £33 000;
York University, Department of Electronics, 1980, £60 000:
Electronics research and development centres

Unlike the Edinburgh region, neither that of Colchester, the home of the University of Essex, nor that of York, where the Univeristy of that name is sited, is highly industrialized. Research groups at both these Universities have, however, established Electronics Centres. That at Essex was founded in 1970, with a Wolfson grant in 1972. Its aim is to encourage the wider use of electronics by firms both inside and outside the electronics industry itself. This Centre has grown slowly, partly because of the reluctance of industry to take up new ideas during a recession, and partly because the industries in the area are widely dispersed. Where a good working relationship has been established, the use of the Centre as the research and development arm of the companies involved has had highly beneficial results. For example, a firm making agricultural machinery which works closely with the Centre has developed a machine for sizing potatoes electronically. The Centre has benefited considerably from the encouragement of the University, and its costs of operation are abnormally low because not all the University costs— such as advice and practical help from academic staff—are charged to the contract work undertaken by the Centre.

The Electronics Centre at York, which was launched at the same time as the University's Department of Electronics, is managed by a former employee of the Essex Centre, who is guided by a management committee, whose members are drawn from all interested University Departments. The multidisciplinary expertise available within the university proved invaluable in devising a technique for solving an inspection problem for a local firm. The Centre also contributes to the teaching and project work of undergrad-uates. Despite the recession the Centre in its first three years tackled both large and small-scale projects for many local firms, including the largest firms in the area. As with the Essex Centre, a period of two to three years elapsed before the York Centre's reputation was established. Word of mouth recommendation has played an essential part in getting the Centre known.

Essex University, Departments of Physics and Engineering, 1980, £39 000, £60 000, and £100 000

Both the York and Essex Electronics Centres have made great efforts to market the services which they have to offer. A second centre was set up at Essex with University funds supplemented by a Wolfson grant. It was based on the Physics Department (£39 000), with the aim of placing contracts with firms, in particular for encouraging the commercial use of computers in small businesses (£60 000). Later the University received a third grant (£100 000) to help it set up an industrial noise unit to develop devices for the suppression or neutralization of excessive noise in industrial premises.

Birmingham University, Department of Electronics and Electrical Engineering, 1980, £110 000: Reliability of electronic components

The Wolfson Semiconductor Laboratory in Birmingham's powerful Department of Electronics and Electrical Engineering provides a service quite different from those of the Wolfson Electronics units already described. In Birmingham the objective is to contribute to the quality and reliability of electronic components. The unit was set up in order to enable a team of solid-state electronics experts to develop, with the help of new equipment, techniques of thermal microscopy and imaging.

Most of the first year of the grant was spent in acquiring and commissioning equipment. Interesting results then followed in quick succession. Semiconductor devices and microcircuits are sensitive to temperature changes caused either by ambient variations or by power dissipation in the circuits themselves. Microelectronic components can be examined under the microscope, and hot-spots due, say, to a defective contact, can be identified by visual or infrared radiation. The Birmingham equipment makes it possible to identify faults either directly or from colour micrographs. Improvements in reliability, design, and manufacturing methods can be expected. The results obtained so far have attracted contract support from several manufacturers. The situation looks promising. There are hopes that income will be sufficient to maintain an industrial service.

Bath University, School of Electrical Engineering, 1980, £127 000: Integrated microelectronic circuits

This grant was made to Bath University for the 'custom design' of large-scale integrated ('LSI') microelectronic circuits. Computer-aided design equipment has been assembled to investigate ways whereby small manufacturers of equipment can profitably produce integrated circuits for their own products in markets too small to attract the interests of the large integrated-circuit manufacturers. For a variety of reasons, mainly financial, collaboration with industry has not been easy, despite widespread goodwill. In the circumstances, the University's own resources have had to be used, where feasible, to augment those of interested companies. It is still too soon to attempt any forecast of the outcome.

A different situation prevails in the same field both at Queen's University, Belfast and at St. Bartholomew's Hospital in London, where units that have been established with the help of Wolfson grants not only provide research and consultancy services to companies, but also concern themselves with the development of 'hardware'.

Queen's University, Belfast, Department of Electrical and Electronic Engineering, 1976, £52 500: Microcomputers and microprocessing

In Belfast Wolfson funds were provided in 1976 to help set up a Signal Processing Unit, in which academic staff co-operate directly with staff specially recruited to help in industrial work. Companies in Northern Ireland requiring the services of a consultant obviously find it more convenient to deal with one of the two Universities in the Province rather than with institutions in Great Britain. Most of the work of the Belfast unit is of local origin, but some contracts have also been obtained from the United States and from England.

A wide range of electronic devices has been selected for development. Among them are microcomputer and microprocessor systems for use in manufacturing, particularly management information systems; a road-texture measurements system for the Northern Ireland Road Services; an electronic suitcase lock; a microwave intruder alarm; and security systems for motor vehicles and for other applications. Prototype equipment has been turned out in the quantities needed for demonstration, and discussions about large-scale production and marketing are continuing. Apart from this development programme, the Queen's University Unit provides consultancy services, and has looked carefully at industrial ventures to which the skills of its own staff, aided by consultants from the University's Department of Electrical and Electronic Engineering, can be profitably applied.

Annual income has grown steadily and reached an estimated £100 000 in 1982. It is now sufficient to maintain a staff of four development engineers and a secretary. There are good prospects for still further growth.

A recent (1983) innovation has been the launch of the Wolfson Unit's 'Microelectronic Broker Service', to help firms in the choice and application of microelectronic devices and components, including custom integrated circuits. The service receives some financial support from the Northern Ireland Industrial Development Board, and the Unit helps to promote the Province as an investment area for the electronics industry.

St. Bartholomew's Hospital, London,
1976, £184 000: Medical electronics

The Wolfson Centre for Medical Electronics at St. Bartholomew's Hospital
was established in 1976. It, too, has identified a number of marketable products which it has carried through to prototype development and for which,
in some cases, it has arranged facilities for production.

Despite the fall in purchasing power of the National Health Service itself,
and the current lack of enthusiasm of companies to support new products,
the Centre has achieved sales of about £120 000 per annum. Even though its
income fell in 1981, it is likely to remain self-supporting.

Nottingham University, Department of Metallurgy,
1968, £255 000: Interfacial technology

A number of grants are listed in Table 3.1 which relate to Materials Technology. Nottingham University received the largest award ever made under
the Technological Projects Scheme in order to set up the Wolfson Institute
for Interfacial Technology. The grant was big enough to provide for a
laboratory building with a working area of 1115 square metres, for the basic
equipment which it needed, and to pay for some staff. From the beginning
the Institute has maintained close links with many University Departments,
particularly Metallurgy and Chemistry. It has also always carried out its
own basic research, much of it with funding from the SERC. Many patents
have been taken out, none of which has as yet generated any revenue.
Exploitation of the patents has been slow for reasons that are discussed in
the following chapter.

In seeking further income, the Institute has transformed itself into a contract research organization, and since the Wolfson grant terminated in
1973, it has succeeded in attracting funds totalling more than a million
pounds. Its work has involved such diverse topics as electrical contact
behaviour, photoconductor characteristics, and many other surface-related
problems including those of oxidation and wear, particularly of materials
used in nuclear reactors. The Institute's income for the next two years is
assured at the level of £200 000. The University has also recently recognized
the quality of work on composite materials by offering to support the
senior member of the team for a limited period. Even so, the Institute's
position, which is dealt with again on page 71, is by no means secure, since
its total income still falls short of the amount necessary to maintain all its
equipment and assets.

Aston University, Department of Metallurgy,
1970, £17 630: Materials consultancy service

Two industrial advisory service units were established at the University of

Aston with the aid of Wolfson grants. The aim of the first was to advise about the choice of materials, usually metals, which would be most effective and most economical for particular industrial applications. For example, an engineering component such as a simple tie-rod, which has to withstand a load of pure tension, can be made either from plain carbon steel or from a stronger alloy steel, in which case it can be more slender. However, the decrease in the volume of the material has to be matched against the higher initial cost per unit weight of the stronger alloy, and the greater difficulty of machining and forming. Design considerations, particularly when more complex stressing is involved, may dictate the choice between two alternatives, or extend the choice to quite different materials, e.g. aluminium alloys, plastics or wood. Each solution will entail costs which not only differ at the start but which vary with time because of wide fluctuations in primary costs and the availability of raw materials, as well as of changes in the cost of energy. Aluminium, for instance, has a density about one-third that of steel, but the highly alloyed steel may be more than three times as strong, so that even in aircraft, where weight is a penalty, there is still scope for using a considerable quantity of steel.

In wartime the substitution of materials is commonplace, being dictated by shortages of raw materials rather than economic factors. The Aston studies have succeeded in establishing suitable methods for assessing the best engineering-cost solutions in specific cases.

Aston University, Department of Metallurgy,
1972, £75 000: Heat treatment of metals

The second grant to Aston made it possible to set up a unit which has become British industry's focal point for information, advice, and education in all aspects of metal heat treatment. Now ten years old and self-supporting, the Wolfson Heat Treatment Centre resembles a small research association. It has more than 300 industrial members who benefit from a computerized information and advisory service that is provided by a team of four full-time staff, who now deal with over a thousand enquiries a year.

Seminars are held to demonstrate and discuss advances in heat treatment methods, and industrial training courses for supervisors are given as required. One such course, 'Understanding Heat Treatment', has been repeated eighteen times in the past six years. Soon after the Wolfson grant was made, the unit also launched an excellent quarterly journal, *Heat Treatment of Metals*, which has continued ever since, with a world-wide circulation. Services to industry have included the compilation of a six-part series, *Guidelines for Safety in Heat Treatment*, and also specifications for the testing of quenching media.

The Centre both encourages and learns from research work on new heat treatment techniques, for example, plasma nitriding, which is carried out in

Aston's Department of Metallurgy. Here the aim of the Centre is not the provision of a consultancy service, but the development of low-cost industrial equipment. This work can be regarded as complementary to that of the Wolfson Plasma Processing Unit, formerly at Liverpool (p. 42), even if in some respects it could be regarded as competitive.

Other heat treatment research has included the development of a computer program to predict the profile of the carbon content in the outer skin of steel carburized in a controlled atmosphere. From such predictions the most economical conditions for the carburizing treatment of any type of steel can be determined. The process can be regulated to ensure minimum distortion and improved performance in service.

Optimization of the heat treatment of tool steels, with particular emphasis on the newer surface treatments—plasma nitriding, vacuum carburizing, and salt-bath treatments producing hard carbide layers has been a major focus of research. The first two treatments improve the fatigue properties substantially, and the third the wear resistance with no adverse effect on fatigue. Modified hardening treatments for high-speed steel giving improved fracture toughness may be used with surface treatments to give much superior combinations of properties.

Some of the unit's basic work on heat treatment has been funded by SERC, and the industrial service has been supported by the Department of Industry.

University of Wales, University College, Cardiff, Department of
Mineral Exploitation, 1968, £10 800; 1970, £64 450;
Department of Microbiology, 1970, £40 000;
University College Industrial Unit, 1980, £54 000

A different type of Wolfson industrial unit was established at University College, Cardiff. The College received grants in 1968 and 1970 for an examination of Welsh mineral reserves other than coal, and was then given a further grant in 1970 to set up a laboratory for the application of microbiology to industrial processes. It had been hoped that enough industrial contracts would be obtained from the start to make the unit self-supporting, but unfortunately sufficient income was not forthcoming. Aided by additional funds from the University, it was therefore decided to establish a Cardiff University Industrial Unit, in order to encourage invention and innovation generally within university departments. The available staff was, however, too small, and it was decided that the Industrial Unit should seek contracts in order to reveal technological fields which held a promise of industrial exploitation, and which could be linked to an appropriate university department. The staff of the Unit are thus middlemen who have to exercise judgement about the likely commercial outcome of a variety of schemes.

Some schemes have been unusual in their nature. For example, a new type of heat-insulating material for protective clothing is being manufactured by a company in South Wales as part of a contract to find new uses for aluminized plastic film. In suitable form this material can replace the usual filling in quilted clothing for a large number of end uses. A window-blind manufacturer, also in the region, is now test-marketing a roller blind which, like the heat-insulating material used in clothing, depends on increased heat reflectance of the material, in this case a new form of laminated plastic. There are other possible applications.

In the field of materials conservation, interest has been shown by Charter Consolidated and by SERC and BTG in the recovery of lead and zinc from steel-works dust. A pilot plant has been set up and full scale operation is likely in the near future. Surplus plastic foam from furniture factories has been incorporated as the main ingredient in light-weight 'growbags', which have been tested satisfactorily for growing tomatoes. A company has been set up to manufacture the bags.

The microbiology group has worked on the treatment of sewage sludge and on the long-standing problem of disposing of farm waste. Both these projects are at the stage of full-scale trials. A further biological project, which stemmed directly from academic research on slugs, has led to a new slug-bait, which the co-operating manufacturer expects to market profitably.

There has been no shortage of ideas at Cardiff, and it is highly probable that the Unit's entrepreneurial approach will generate several profitable new products and processes. The Wolfson grant of 1980 has been used to increase the number of project managers, who act as links between the University and industry, and who work alongside the academic staff of the College. They were originally engaged on renewable annual contracts, but with the funds now available from the Foundation, they can be given greater assurance of continuity of tenure. The Manager of the University Industrial Unit has the full backing of the Principal of the College (who acts as Chairman of the Management Committee, and thus provides valuable support to the Unit which he himself set up). The Unit now derives its income both from royalty payments and from reserves built up over the years. It is noteworthy for the attention it gives to helping small industry in the Cardiff area, and for its diligent search for opportunities in the market place.

5

SUCCESS AND FAILURE

By the time a Wolfson technological grant comes to an end, the recipient, given that he has succeeded in his aims, can expect to be earning enough for the unit he has set up to be making profits or, at the least, to be self-supporting. Since the grants are, as a rule, not renewable, if he had failed he would either have to look elsewhere for funds to keep his work going or bring it to an end. In practice more than 70 per cent of the projects which the Foundation has supported have continued to function when the grants ran out, a rate of survival which has greatly exceeded the expectations of the Trustees when the scheme was launched in 1968. Some illustrative case-histories, and a few tentative views about the reasons why some projects have proved more successful than others, are discussed in this chapter.

Diverse university-industry relationships have developed under the scheme. They include university departmental arrangements to instruct companies in new technology (i.e. 'technology transfer'); the provision of research services for industry, either inside or in close connection with university departments; the establishment of independent research services; the establishment of university companies; and finally, the exploitation of the results of university research by an outside company.

CASE STUDIES (A):
RESEARCH, TEACHING, AND CONSULTANCY SERVICES

University of Manchester, Department of Engineering,
1980, £190 000: An industrial unit in maintenance-related engineering

Terotechnology—the study of the installation, maintenance, and use of machinery— was an interest of the engineering laboratories of the University, and the Wolfson grant was intended to encourage enquiries into specific matters of industrial concern. Although terotechnology can be studied theoretically, and while teaching can obviously deal with general principles, the educational benefits of focusing on the problems of a particular industry, taking into account its management and financial circumstances, are obvious.

The Manchester Wolfson Unit was planned (i) to provide for commercially based short courses on all aspects of plant maintenance; and (ii) to operate a commercial advisory service on 'condition-based maintenance', by which is meant that it may be more costly to carry out regular maintenance at pre-set intervals than by reference to the actual state of the parts of a plant. For instance, the servicing of a motor car every 15 000 miles, with attention being paid between services to any parts that show signs of wear, may be a more economical way of keeping the car in good condition than routine servicing at 7 000 or 10 000-mile intervals. To decide which regime is the best in a manufacturing plant, data have to be collected, assembled, and analysed. An advisory service must therefore be ready to provide (i) day-to-day monitoring services; (ii) training of company staff to take over 'condition' monitoring; (iii) development of condition-based maintenance techniques; (iv) consulting work on condition-based maintenance and monitoring; and (v) advice on policy and the analysis of data. Each of these activities is of interest in undergraduate, postgraduate, and post-experience education.

Determining the optimal requirements for the routine maintenance of motor vehicles is a simplified example of the work of the Manchester Unit. It is much more difficult to calculate the most economic maintenance costs in heavy industries such as electric-power generation, steel-making, ship-building, railways, and oil production and refining, industries which are massively capitalized and where plant may be in continuous operation. Many of the devices needed for condition-monitoring have yet to be developed.

At least as important is the difficulty of persuading a large heavy-engineering company that a university department can be of help to them, and that it is worth while to spend money on the relevant research programmes. As a rule links will be forged only when the managers of heavy-engineering firms are themselves interested in the study of their own maintenance procedures.

Because it had been found difficult to make contact with large companies at the appropriate level, the progress of the Manchester Unit was bound to be slow. But the combination of teaching, research, and consultancy which it undertakes is attractive, and during its first years of operation the Unit has managed to arrange some modest study-contracts with British Steel, as well as with several other big companies. A working relationship has also been established with Yugoslav Railways for the condition-monitoring of its vehicles, and a number of other overseas contracts have been obtained. Some oil companies have now shown an interest in the Unit's work.

Income has grown only slowly, but the prospects look good, with demand from overseas, both for consultancy and for fully trained maintenance engineers, increasing. Suitable staff have to be recruited to replace those who are attracted away, and the Unit has now become a training centre for maintenance engineers. Sales of software systems to aid the recording and processing of condition-based maintenance data, the development of which has been under way for two years, have started and could also prove profitable. Revenues should reach £100 000 in the year 1983/84, chiefly from training courses and consultancy contracts. This will be enough to meet the Unit's present day-to-day costs.

In some respects the history of this Manchester maintenance-engineering Unit is not dissimiliar from that of the Edinburgh Microelectronics Institute in its first years. The industrial environment of Manchester is, however, very different from that of Edinburgh. The more prominent industrial concerns in the Manchester area, for example British Steel and the fuel industries, are essentially regional offshoots of their parent companies. They seem less interested in establishing links with university departments than are the electronic companies in Scotland, many of which are small, while others are branches of large electronic companies. All appreciate the value of the University industrial unit as a source of information about the latest developments in their field. The Wolfson Units at Belfast and Cardiff also find it easier than the one at Manchester to market their wares and to form links with local industry. The Manchester Unit has thus had to look further afield, as it has done with Yugoslav Railways. Whether it can develop as part of the University, or is driven to become more of an independent consultancy and training service, remains to be seen.

It would be fair to say that one difficulty with which the Unit has had to contend derives from its very success in pointing the need for trained maintenance engineers, and as a consequence losing its own trained staff.

University of Stirling, Department of Biology,
1976, £68 400; 1978, £37 800; Establishment of a Unit of Seed Technology

The purpose of the first grant made in 1976 was the establishment of a research unit whose work would be related to the needs of seed producers. The initial objective of the Unit was to find out why certain seeds with a high rate of germination in laboratory tests often have a poor emergence rate when sown in the field, particularly in adverse soil conditions. It was soon discovered that when the seeds are soaked in water, certain chemicals, for example potassium, are leached into the water, whose electrical conductivity rises owing to the presence of the leached ions. Batches of seed that are unsuitable for early sowing can therefore be identified by a conductivity test. A simple test of this kind can be of great economic value to the grower, for not only does it save the immediate cost of the seeds and the losses resulting from poor yields, it also obviates the waste of time and effort that would be needed for resowing.

The conductivity test also provides a means of detecting other causes of low vigour in seed production. For example, cracks in the outer coat of the seed, which may occur either in handling or because of maltreatment during storage, may encourage a rapid intake of water. The test requires only elementary apparatus and little technical skill. A second test applicable to small-seeded vegetables was also devised and developed by the Unit. This so-called 'controlled deterioration test' again is technically straightforward.

These uncomplicated test procedures were developed by the Unit in order to facilitate the transfer of knowledge to an industry that had little tradition of scientifically based methods. They are now used routinely by seed companies both in the UK and abroad. They are also recommended by the International Seed Testing Association, which oversees the testing of crop seeds in international trade. Paradoxically, however, the very simplicity of the tests, which could be readily carried out by the seed producers themselves, militated against the economic success of the Unit. The producers could not be persuaded to finance further research work, nor were the tests of a kind from which royalties could be derived. The unit has, however, earned some income from training courses for the staff of seed companies. Contracts have also been obtained for the study of the effects of agrochemicals. Another line of research which is being pursued is in the non-destructive viability of crop seeds.

Wolfson Microelectronics Institute, Edinburgh University
(see pp. 58-9)

The Edinburgh Institute, which was set up in 1968 with a grant of £130 700, has been self-supporting since 1974, with contract income covering direct costs. The University first provided lighting, heating, telephone, rates, and financial services, in exchange for the educational work that the Institute does for the University. More recently this arrangement has been changed, with the University making a charge for its services to the Institute. The Institute became self-sufficient more rapidly than did the one at Nottingham, with income growing steadily from £60 000 in 1974 to about £500 000 in 1981/82. Outstanding debts to the University were repaid in 1981, and in 1982 the surplus on earned income was spent on replacing and extending equipment.

The founders of the Edinburgh Institute were members of the University Department of Electrical Engineering, a Department which already enjoyed a high reputation for the quality of its enquiries into light-current and electronic engineering. Several ideas that had emerged in the course of the Department's work appeared to have industrial potential, and it was hoped that their exploitation could be furthered by the creation of a unit which would also have an educational role, by revealing to students the intellectual challenge of work in industry. In the early stages, staff attached to the Institute helped supervise both undergraduate and postgraduate students. The location of the Institute alongside the Department responsible for its foundation made such educational collaboration mutually convenient.

Financial pressure on the Institute grew as the initial Wolfson grant was expended, and it became necessary to place tighter controls on expenditure, on the use of equipment, and on staff time. Academic and research staff of the Department, however, continued to serve the Institute as part-time consultants, thus keeping undergraduates in touch both with industrial needs and with technical advances in a rapidly developing field. The initial Wolfson grant was not intended to provide more than modest and temporary accommodation for the unit, and it was not until 1978, when the Institute was clearly self-supporting, that a specially designed building of over 900 square metres was provided with the help of independent local finance.

The Institute operates in many ways as if it were an industrial company, of which the Director of the Institute is, in effect, Managing Director. Interposed between the Director and the University is an advisory board appointed by the University Court, and consisting of approximately equal numbers of university and industrial members, under the chairmanship of the head of the Department of Electrical Engineering. Within the Institute, day-to-day management is in the hands of the Director and his executives. A full-time financial controller sees to it that accounts are kept separately for each project so as to assure full recovery of overheads. The Institute's financial accounts are presented monthly on normal company lines. It is interesting that the Institute has chosen to run its affairs as though it were a company, without actually being registered under the Companies Act.

In their efforts to revitalize Scottish industry, the Scottish Office, the Scottish Development Agency, and the Scottish Economic Planning Department have done much to encourage the growth of the electronics industry, with the result that the central belt of Scotland is now the home of several microelectronic companies, both British and foreign. It is not the responsibility of these Government agencies to support research in universities, and since the Microelectronics Institute is part of Edinburgh University, its dealings with the agencies, while cordial, were not always helpful.

In order to develop its services to industry, the Institute has always drawn upon the specialist interests of the University's Department of Electrical Engineering. More recently, it has developed its own research programme, which now accounts for 11 per cent of the Institute's total turnover. It has designed integrated circuit techniques to meet the needs of individual customers in devices such as microprocessors. Another of the Institute's achievements, developed in collaboration with the University Department of Computer Science, is the GAELIC (Graphic Aided Engineering Layout of Integrated Circuits) software design suite, now marketed by Compeda, a subsidiary of the British Technology Group. Other saleable products are a control chip for an electronic micrometer; a helium speech-unscrambler for deep-sea divers; a microprocessor for controlling a heart-rate monitor; secure communications equipment for fishery use; and an 'intelligent' drinks dispenser. Some of these projects have taken many years to perfect. The GAELIC software project, which was carried out with the help of graduates recruited from industry, took eight years.

The design and application of integrated circuits is now a valuable source of royalty income, and the Institute's success was, no doubt, a major factor in the SERC's decision in 1979 to establish a centralized fabrication facility in the Department of Electrical Engineering.

The move into the design and manufacture of specialized products is leading to the establishment of new companies. Two former members of the Institute have set up one to deal with integrated circuit designs. Another offshoot company is embarking upon the production of safe communications and has started to sell a special-purpose keyboard which simplifies word processing on standard microcomputer systems. This is the first product wholly conceived within the Institute and carried through market exploration to sales. It represents the end of a development chain, from research through service activities and royalty income to final production and sales.

The foundation of a chair in Microelectronics by the Lothian Regional Council, noted in Chapter 4, is not only a further indication of the regard in which the arrangements in Edinburgh are held, but also ensures that the educational objectives of the Institute will not be overlooked.

All the new electronics companies that have been set up in Scotland are well informed about the field of technology in which they are working. To further the application of micro-electronics in small businesses, a company called Integrated Micro-applications Ltd. (INMAP), was formed by Edinburgh and Heriot-Watt Universities in 1981, work being sub-contracted, as appropriate, to the Edinburgh University Wolfson Institute, or to suitable research groups in Heriot-Watt University.

The Edinburgh Institute based its service to industry on the research work of the parent department in the University, and did not in its early years attempt a substantial research programme of its own. When it became self-supporting, it began research work in its own right to further its industrial research, but it also began to develop devices for which a sale could be forecast and received a further grant in 1981 to further this work. The interest in products for manufacture could well have arisen from a feeling for commercial ventures derived from the industrial form of the Institute and the reaction to this in members of staff.

Institute for Interfacial Technology, Nottingham University
(see p. 62)

In 1982 Nottingham University made a submission to the House of Lords Select Committee on Science and Technology,[8] in which it gave an account of some of the problems that had been encountered after the Institute's formation in 1968 with the help of a Wolfson grant. The then Vice-Chancellor had described the Institute's aims as being directed towards undertaking 'basic research focused by specific long-term technological concepts rather than by general investigation of a broad scientific area'. Its stated objective was to provide 'a strong interaction between the university and industry and to make available experimental facilities and equipment on a scale that could not be matched by a single industrial firm. By taking part in this research, younger academic scientists would be orientated towards industrially useful research and dvelopment and become aware of industry's problems and needs.'

These clearly stated aims failed to attract income from industry on a scale that would maintain the Institute—once the Wolfson grant had expired—in the way that had been hoped for. The objectives were therefore re-specified some years later as follows: 'To promote the application of basic science to the development of new processes and products, by (a) providing a professional service devoted to the investigation of problems encountered in the production environment, and (b) the use of earnings to initiate and subsidise in-house research and development programmes with a substantial innovative content and commercial significance.'

Most of the research that has to be carried out in the development of new products and processes needs to be related specifically to the requirements of individual companies, and to their production and marketing plans, information about which will not be publicly available. To sell its services, the Institute therefore had to establish close and confidential contacts with potential customers and suitable industrial collaborators. This is something that almost all university industrial units have to do, but the problems are considerably eased when a university is situated in an industrially developed area, and when the university's skills can be related to the interests of firms which are not too far afield. The East Midlands region is the home of diverse manufacturing concerns to which the particular capabilities of the Nottingham Institute would, it was thought, be relevant.

The original assumption that small firms would be the most likely clients proved, however, to be wrong. Larger and medium-sized companies were not only more interested in collaborative work, but were also prepared to recognize the advantages of utilizing skills and equipment which they themselves did not possess, and to pay the 'going rate' for these costly services. Companies which employ highly qualified scientists and engineers find it easier to judge when a university unit's ideas and techniques may be of value. Even a very large firm with a strong research department will often find a need to turn to the scientific departments of a university, in which work is going on at the frontiers of knowledge in a wide range of disciplines.

To earn income, the Nottingham Institute undertook various forms of contract work, ranging from research and development programmes lasting several weeks, to brief investigations involving no more than specialized measurements or chemical analysis using equipment that was not available to the client. Straightforward chemical analyses are frequently commissioned for this reason. The Nuclear Installations Inspectorate, which has no in-house capability for research and development or for testing, has also been one of the Nottingham Institute's regular clients. Surface properties of materials and surface reactions affect the efficiency and safety of nuclear reactors, especially the British AGR and Magnox systems, and fundamental knowledge in this field has to be available not only to the Inspectorate, but also to the CEGB and the UKAEA, both of which are concerned with nuclear reactor design. Since the Nottingham Institute was founded as a centre for the advanced study of surface phenomena, it was not surprising that support from all these bodies was forthcoming from the outset. It is only to be expected that the public bodies such as the Nuclear Installations Inspectorate should find merit in turning to an impartial university institution for enquiries, the results of which often become public and controversial.

If they are to attract contract work, university/industrial institutes, such as that at Nottingham, need to maintain a programme of basic research of high quality either within the institute itself or in related university departments—ideally in both. Moreover, a reputation as a centre of excellence can be sustained only if sufficient income can be attracted to support and extend the research base. The cost of basic research should therefore be treated as an overhead which the institute recovers in its charges to customers, many of whom do not appreciate that first-class research is expensive. That was not the case with Rolls Royce Ltd., which provided the Nottingham Institute in its early days with the support necessary to embark upon speculative research with the aim of producing new types of damage-tolerant, fatigue-resistant composite materials. The concepts underlying this work were original, and became the subject of patents taken out by the NRDC. Funds for continuation of the work have been provided by the SERC, NRDC, and the Institute itself. But the recession of the past few years has made it difficult for the Nottingham Institute to attract sufficient industrial support for the development of new products.

When in 1980 the Foundation suggested that applications for project grants might be directed towards 'R and D projects related to any aspect of UK industrial activity that could lead to a new product or process development', the Nottingham Institute succeeded in obtaining a grant for improving the characteristics of photoconductors intended for use in the latest generation of copying machines. This speculative project was closely related to areas of research that have been pursued by the Institute since its foundation. The project has every hope of being technically successful.

Although much excellent work has been done at Nottingham, it cannot be claimed that the Institute has been wholly successful and, indeed, the submission to the House of Lords indicates that there are still difficulties to be overcome before its future can be assured.

CASE STUDIES (B): UNIVERSITY COMPANIES

Three universities in which new products or processes have been developed—

Queen Mary College, Salford, and City—have formed limited companies which operate under strict industrial discipline. At Queen Mary College, the company, QMC Industrial Research Ltd. (QMCIRL), sets up subsidiary companies to deal with new activities; at Salford, subsidiary profit centres are set up by the main company, Salford University Industrial Centre Ltd.; while City University has established a company, City Technology Ltd., to market a specific device.

QMCIRL has dealt with three Wolfson grants. The first, in 1974, provided for linear programming studies (p. 40), originally in relation to the paper-making industry. Little revenue resulted from this work, the results of which have been applied elsewhere. The benefit of the range of services provided by the parent company were obviously valuable in the period of transition from one sales area to another.

It is unlikely that the same problems will arise over the 1981 grant for work on the suppression of smoke and toxic gases from burning polymer foams. The subsidiary company formed to promote this work will be responsible for the commercial exploitation of such products and processes as might result. The related advisory or consulting services for industry should produce quicker returns and reduce any cash-flow difficulties.

The most difficult project for exploitation will be the Queen Mary College anchors (p. 22). QMC Anchor Technology Ltd. is responsible for licensing and royalty arrangements with manufacturers, at the same time as research and development on anchors remains under the direct control of the academic staff who were responsible for the original design concept, but who no longer have to spend time dealing with financial, legal, and manufacturing problems. The establishment of an anchor design team within the company, working with the original inventors of the concept within the College, is particularly appropriate when exploitation entails a jobbing form of manufacture.

The Salford University Industrial Centre Ltd. is a single company which oversees a variety of interests which it manages by way of separate sections of the main company rather than through subsidiaries. In accordance with modern management practice, each section has to justify itself by the criterion of profitability. The section of SUIC that deals with ion-plating, referred to on page 43, arranges for routine plating as required by customers. It also designs new equipment which clients may purchase for their own factories, drawing on the knowledge and experience of colleagues in the University Departments of Aeronautical and Mechanical Engineering and Electrical and Electronic Engineering. Both these Departments carry out general research on plating, and experiment with hard coatings by ion-implantation. Facilities for precise machining are needed for this service but, unlike the QMC anchor product, heavy engineering equipment is not called for. Six full-time staff of the university company deal with most of the *ad hoc* research, development, design, and manufacture of ion-plating equipment. This constitutes a growing business and may well form the basis of a viable independent company. SUIC has had to compete with established firms at a time when business has been in decline.

It is only to be expected that a new university industrial company will find that expenditure on salaries, wages, stores, and equipment at first exceeds income, with a consequent negative cash-flow. This is well illustrated by the experience of the Wolfson Unit for Electrochemical Technology at City University. Its origins lie in research which was carried out in the University jointly with the Electrical Research Association (ERA). This work encouraged the Ministry of Defence in 1972 to make a grant of £16 000 for the development of a gas-sensor warning system for the protection of air crews against failure in their supply of oxygen. In the same year City University also won a Wolfson Technological Projects grant of £36 050. This helped keep the unit alive when, three years later, the defence budget came under pressure, and the defence grant was cut. Efforts were then made to launch the oxygen sensor commercially, and the research programme was kept going with the help of support from the National Coal Board.

The Wolfson grant to City University was not tied to a precisely specified research programme, and was available for the development of equipment which had been identified as an industrial requirement. For example, there was a clear need for a device to detect shortages of atmospheric oxygen in the sewers. The design, development and field trials of such an instrument were successfully carried out by the Wolfson Unit in co-operation with the ERA and a NRDC company. It was then decided, however, that since the University had most of the specialized knowledge, a new company, City Technology Ltd. (CTL), should be formed to take over all the rights and responsibilities of the University's partners, the ERA position being safeguarded by the assignment, under a royalty agreement, of the patents and other rights to CTL.

Statutorily City University's Wolfson Unit and the University's company are separate bodies. The Unit is an arm of the University which is concerned with a particular field of study; the company is a commercial entity. The two are, however, closely linked functionally, with the unit continuing to develop such gas-sensing devices as the company judges to have commercial value. As in any university department today, the unit is at liberty to seek contract income—for example, as it has done from the NCB for work on the detection of carbon monoxide and, more recently, of oxides of nitrogen. Sensors for hydrogen sulphide, sulphur oxides, chlorine, and carbon dioxide are also attracting attention. In 1981 the Wolfson Foundation therefore made a grant to extend the Unit's work. But since CTL is now making good profits, in future it will be expected to cover its own development costs. Where a special interest exists—as in the case of the NCB's concern about carbon monoxide—the client might well be expected to contribute to the cost of the work.

The oxygen sensor has been marketed successfully but sales only became appreciable in 1977. Even in that year, and with additional revenues from research contracts, income failed to meet outgoings. None the less, a balance of cash from previous years' grants and contracts was sufficient to meet the shortfall. In 1978 cash-flow remained negative. In 1979 it remained negative but had ceased to fall. This was the real turning point. In 1979/80 sales were about £160 000, and in 1980/81 £310 000. They are still rising.

The successful way that City University has managed its affairs provides a useful lesson, but one which may not be applicable generally. City Technology Ltd. has thrived because the individuals who started it were able to adapt readily to changing circumstances. Their success was recognized by the conferment in April 1982 of a Queen's Award for Technological Achievement.

Although the grant to York University in 1978 was not developed through a university company, the external company which has been responsible for exploitation has concentrated, like City Technology Ltd., on a prescribed range of products. The market assessment and the definition of the likely saleable product was made outside the unit and some of the problems of development, manufacture, and sale are not dissimilar from those of City Technology Ltd. There would seem to be advantages when an active technologist with market knowledge and considerable enthusiasm enlists the aid of a university. The company, as recorded in Chapter 3, gained a Queen's Award and appears to be highly profitable. One problem which needs further study is how to discover more opportunities of the same kind.

GENERAL CONCLUSIONS

While it is all but impossible to say why some projects have prospered so much better than others, it is obvious that no university team which sets out to sell its wares or its expertise either to a manufacturing firm or to an entrepreneur can expect to succeed if it does not understand the ways of the market. Nor could any commercial company be expected to co-operate with a university department unless it were convinced that it could make practical use of the results of the department's research. A Wolfson Technological Project must therefore have both commercial and scientific or technological objectives in view, and those responsible must be ready to help market their products or services. They must also recognize the need to meet the undertakings they make at the appointed times. The interests of a unit or institute may also have to be widened, in a way that was not originally intended, in order to acquire income or attract a great range of industrial custom.

Obviously a university technological-project unit needs the help of enthusiastic and sympathetic industrial partners from the very start. This, unfortunately, has been lacking in some cases. While it has usually been easier to form links with local and frequently small, companies, large firms farther afield have often proved more responsive. The highly-qualified scientists and engineers whom they employ can more readily appreciate what the universities have on offer than can the staff of small companies.

To succeed, however, a university/industrial unit or institute clearly needs more than a saleable service or product. Unlike an academic research team, it also has to establish its 'credibility' with prospective customers, who need to be assured that the unit can keep to whatever timetable it may agree with clients for the work it undertakes. The unit also needs to keep up its sales drive, as do Research Associations and commercially sponsored research laboratories. With few exceptions, this has been the most difficult part of the exercise for those universities that have fostered Wolfson units.

The advantages of a technological projects unit appointing an industrial liaison officer or sales agent to assess the market and to foster relations with potential customers are obvious. Such a staff member can keep academic research workers in touch with a company's technical staff and arrange staff exchanges, in the same way as a company may find it profitable to set up its own organization to monitor academic developments and to spot potentially useful research.

On occasion a Wolfson unit or institute has singled out a new product or service for special promotion, and has then been bombarded with an embarrassing number of enquiries. Where this indicates the existence of an untapped market, it should be possible for the institute to develop wider services to industry. But when marketing, financial control, and the management of resources assume increasing importance as business expands, a structure like that of a company may well prove to be the most effective way of underpinning the unit's operations.

It might prove even more helpful to establish a registered company to exploit the results of a unit's work, particularly when these take the form of specific products or processes. As well as benefiting from the opportunity to learn about company law, consultancy contracts, royalty payments, and patent rights, the academic members of a project team will then have more time for their research, leaving other members of the company the responsibility of finding industrial collaborators and of arranging consultancy and design work for customers. If a company or a Wolfson unit or institute is to be of value to industry, it should always aim to maintain the closest possible links with the university to which it belongs. It is this close contact with back-up research within the academic departments which distinguishes these organizations from a sponsored research organization or a research association.

When a university industrial unit has succeeded in developing several different products or processes, each suitable for exploitation, the university concerned may find it advantageous to form a holding company, such as that set up by Queen Mary College, and to make each product the concern of a subsidiary company, whose continued existence would be dependent on its profitability. The holding company would retain overall responsibility for accounting, patent protection, and agreements with industrial collaborators. Alternatively, as is the practice at Salford, each product could become the responsibility of a division of a single university company, each responsible for making a profit on its transactions. Which of the two arrangements is the better would depend on the size and variety of the marketable products.

Even when a university may not find it appropriate to set up a registered company, technological units or institutes under its wing could benefit from modelling themselves on the structure of a company. Doing so would ensure strong executive control and proper financial accounting for each project, with provision for the recovery of overhead charges. Overall responsibility should rest on a director and a financial controller, advised by a board consisting of both academic and industrial members. The inclusion of academic members is most important to maintain a formal link between the unit or institute and the relevant university departments. Without such a link, a manufacturing activity will tend to drift away from the university and university support will gradually be lost.

When an industrial institute or a university industrial company itself undertakes to manufacture part of a product or to provide a service for a firm, the scale of the facilities and equipment that would be required should never exceed what can be readily accommodated within the confines of the university. Some Wolfson technological project units that started to invest in specialized testing equipment and production facilities have failed because of cash-flow problems that arose before the unit started to generate income.

The reverse, as has already been explained, can also happen. The commercial collaboration with the Stirling seed project led more quickly than had been expected to a non-patentable process which satisfied the needs of interested companies—which then withdrew their support. The university unit had then to start working for patentable 'intellectual property' from which to derive revenues. But even when the results of high-quality research can be protected by patent, income from royalties may still be slow in coming without determined efforts in marketing.

It is not surprising that unpredictable or extraneous factors may make it necessary to redirect the primary aim of a technological project team. The recent downturn in the British economy has been a major spur to such changes. A unit that started with a limited goal may have had to evolve into one with more general objectives, while a unit set up to develop a new product or process may have found it appropriate to become an advisory unit. But whatever happens, if income is to be attracted from industry, it is essential that technological project units remain in the forefront of their fields of specialized knowledge. They need to be recognized as 'centres of excellence'. They either have to carry out relevant basic research themselves, or keep in the closest touch with academic colleagues whose researches are relevant to the work of the unit. In accounting terms, the cost of basic research, which will usually be of little immediate short-term interest to industry—even if it may have ultimate long-term value—should be regarded as an overhead to be recovered in general contract income. Industrialists usually regard the education and training of university graduates as a charge on general taxation. When a Wolfson unit provides an educational service to its university, either by teaching undergraduates or by supervising postgraduate students, the university may in return cover the unit's overheads, such as the cost of lighting, heating, telephone, and rates. Such educational activities have the secondary benefit of making both students and members of staff aware of industrial problems and opportunities.

These generalities, derived as they are from the history of the Wolfson Technological Projects Scheme, are little more than might have been expected. No projects were supported when it was judged at the outset that they were unlikely to be of commercial interest. Obviously some of the applications that were rejected might have succeeded where others that were supported failed. Unfortunately, however, the record of the scheme so far does not provide any more certain a basis for judging the merits of individual proposals than the process of informed assessment that has been pursued so far. Some projects which were supported derived from new scientific knowledge, whose practical and industrial worth the research workers concerned were themselves able to appreciate. Other proposals derived from academic research which had been directed along a specific path related to an obvious industrial and commercial need. In either case, success depended on a mutual and intimate awareness by the university team and its industrial partner of their respective capabilities and interests.

It has been a common experience that those in charge of projects which have been awarded grants under the scheme have understated the time it would take to achieve their objectives. Academic scientists and engineers do not easily adapt to the exigencies of commercial life. But contract work has to be fitted into a timescale, and academics need to appreciate that a greater sense of urgency applies to the solution of industrial problems than necessarily does to that of a problem in a field of basic research. A commercial or quasi-commercial unit enclosed within the teaching and research institution in which it was created, and to whose ways it is accustomed, has to learn to change the manner of its operations. Exchanges of staff between the university group and the firm with which it is collaborating can obviously help in making academics understand the realities of business life.

This is one reason for feeling that an organization for helping industry should be outside the university departmental structure, since many staff in Wolfson groups have conditions of service more closely related to industry, for example, lack of tenure, than to university practices. Such staff may well see their avenue for promotion outside the universities, and although this may temporarily weaken the Wolfson groups, it is probably no bad thing. Even so, the conditions of service need to have the situation with a university in mind if good people are to be retained in the units.

By its very nature, a charitable foundation cannot do more than launch a new university industrial unit, or assist one which is already at work. It has been the experience of many who have been supported that their Wolfson grants have run out before they have become self-supporting, making it necessary for them to seek financial help elsewhere. In case there should be either a need to augment the grant from the Foundation or to prolong the research when the grant has expired, Wolfson project leaders would therefore be well-advised to inform not only the Foundation, but also other grant-giving bodies, for instance the BTG and SERC, about the progress of their work. Government departments and other public agencies should also be made aware that a university institute's work is directly relevant to a local industry which it is in the public interest to foster.

The interests of regional and local authorities and organizations in the work of Wolfson units can be more important. The governing bodies of most universities include many representatives of local communities who can help in furthering the progress of a university industrial unit. Among those who have benefited directly in this way are the team at Liverpool University, which was helped by a Wolfson grant in a co-operative venture with the Polytechnic and the Merseyside County Council to carry out research studies related to developments which the County Council wished to foster. Both the Edinburgh Microelectronics Institute and the Salford University Industrial Centre have also worked closely with local authorities. The University of Manchester and UMIST have also received grants to further work in support of local textile industry.

Every bit as important as the encouragement which local authorities and organizations can give to the work of Wolfson technological units, perhaps even more important, is the continuing interest of universities in the welfare of the units in whose formation they played a part. Sometimes they have stepped in to provide support when the term of the Foundation's grant has expired. They have usually been as sympathetic as has the Foundation to changes which extraneous factors may have dictated in the objectives of a unit. Indeed, without the help which parent universities have themselves provided, it is all but certain that the proportion of the Wolfson projects which have survived would have been far smaller than has proved to be the case.

The Wolfson Scheme from its beginnings in 1968 has made grants in accordance with the procedures described in Chapter 2, and these leave the responsibility for determining the type of application with the universities. When an award has been made, the university decides how it will be administered, whether within a university department, a separate unit, a company, or in any other way, and the choice and form depends to a degree on the aims and the local circumstances. When in a few cases these change, the particular university has generally consulted the Foundation on changes in programme or procedure. The changes would be discussed, but never has the university view been overruled so that the full responsibility for success or failure lies with the universities.

Universities know that they are the main founts of the scientific and engineering knowledge from which technological innovation springs. Even if not explicitly stated, it is one of their responsibilities to maintain and improve the technological standards of the country. While it is not the purpose of the Wolfson scheme of technological grants to divert universities from their traditional objectives of teaching and research, it is the Foundation's continuing hope that universities and technological colleges will help to find even better ways of transferring to industry new and useful knowledge more quickly than is still the case. While the industrial application of new ideas has to take place through the medium of individual companies, on whose profitability the health of the economy depends, experience indicates all too clearly that one of the most difficult steps in the process of technological innovation is to decide what knowledge is potentially exploitable to economic advantage. This is where the Foundation has tried to help.

POSTSCRIPT

Lord Zuckerman, OM, KCB, FRS

The economic climate of the United Kingdom has changed dramatically since the Wolfson Technological Projects Scheme was launched. What has not changed is our dependence on manufacturing industry for the goods that we must export in order to help pay for the food and raw materials we buy in from abroad. Despite the considerable fall in the UK's share of world trade over recent decades, manufacturing industry still provides about a quarter of our GDP. It is responsible for three-quarters of our visible trade exports. And in a period of high unemployment it still accounts for about a quarter of the nation's jobs. A thriving manufacturing economy remains critical to our national well-being.

Today there can be no industry which is not to greater or lesser extent science-based. Our so-called 'high-tech' companies—for example those which are involved in what is becoming known as the 'information-technology explosion'—cannot hope to survive if they are not as well informed as are our competitors about the new knowledge that emerges from the basic research carried out in universities. Our primary producers—whether of food, coal or steel—as well as all our traditional industries, need every advantage that scientific knowledge can bring to bear if they are to hold a place in highly competitive markets, where they have to contend not only with the industrial and commercial skills of other manufacturing countries, but with distortions of trade due to inequalities in the costs of production and in the price set for the ultimate product. To keep up to date, our industries cannot do without the universities, and the latter are far more aware now than they were fifteen to twenty years ago of the value of the 'intellectual property' they have to sell.

As explained in Chapters 1 and 2, the Wolfson Scheme was designed to help narrow the traditional gap between academics and industrialists. That it has succeeded is abundantly clear from the preceding three chapters. Indeed, as has already been said, what has been achieved exceeds by far what was hoped for by the Trustees when the Scheme was launched. The universities have benefited from the establishment of more than a hundred industrial units which are now self-supporting. They have profited because the latter have generated for their parent institutions far more than the £17 million that the Foundation has expended on the Scheme. Jobs have been created for university teachers and specialized staff. Students and academic research workers have been helped to realize that research can be both intellectually satisfying, and capable of being steered to a practical and financially rewarding end. Industry, and particularly small local companies, have been made aware that the new knowledge of which universities are the fount can be turned to practical ends and profit.

This, of course, is not news to Britain's bigger companies. For years they have maintained close contacts with the science and engineering departments of our universities, from which they recruit technical staff, and with which many have always contracted for specific researches. ICI, for example, claims that it maintains 'a wider range of better research contacts with universities than any other company in Europe'.[9] What is more, it is usually the big and well-capitalized companies that have the resources which are needed to be first in the exploitation of new technological ideas. Be that as it may, it is none the less the smaller company, whether established or new, that has usually benefited most from the Wolfson Scheme, as it will have done generally from the direct university/industry links which are now also being encouraged, even if not in exactly the same way, by both the UGC and the SERC, as well as by the Department of Industry.

The Wolfson Scheme is widely acknowledged as having helped bring about what is undoubtedly a fundamental change in the attitude of governmental institutions to the establishment of commercial links between universities and industry. Apart from the funds which the Government now makes available for this purpose, amounting to some £20 million (if the industrial studentship (CASE awards) and the Teaching Company Schemes are included), it also spends about £2100 million on research and development contracts in industry (about half of it by the Ministry of Defence[10]), of which a small fraction is subcontracted for work in universities. In addition, some £600 million has been allocated in the 1983 budget for the loan guarantee scheme for small businesses. Fiscal measures also encourage research and development within industry, although to a much lesser extent than in the United States.

A recent ACARD/ABRC Report,[11] issued at the Prime Minister's request, to the title 'Improving Research Links between Higher Education and Industry', commends the Wolfson Scheme. In a separate paper,[12] the Chairmen of these two bodies now propose that the Government should establish an 'industrial seedcorn fund' for the further encouragement of closer links between universities and industry. The size of the fund should, it is suggested, equal 25 per cent of all the monies earned by Higher Education Institutions 'through contracts, consultancies and investigations from the private sector and the public trading sector' (which they now estimate to be about £40 million p.a.).

Given that this recommendation is accepted, it will be necessary to devise a procedure for the selection of the projects which are to be supported. The two Chairmen recognize that this will not be easy, and they point to the danger of carrying selectivity to the point where some avenues of enquiry may be blocked. They correctly warn that 'the choice of priorities in basic science carries an inevitable risk of error—which may not be evident for some years'. Selection in the field of applied research and development carries more immediate risks. In today's competitive industrial world, errors in selection can end in financial disaster for those concerned.

It would, however, be naive to suppose that it will ever be possible to evade the element of risk which characterizes every step in the lengthy process of innovation—a process that starts with applied research, and which, given success, ends in marketing and sales. The engineer or applied scientist who believes that something of practical value can be made of a new piece of 'basic' knowledge as often as not proves to have been over-optimistic. The development engineers who design the plant for a new product or process usually underestimate the difficulties. Those who do the market research not infrequently reach the wrong conclusions. Mistakes cost time and money, and in the case of major developments, tens of millions of pounds have often been lost. Knowledge and judgment of the highest order are called for in deciding when to press on with a particular project, or when to cry 'enough is enough', to admit that it is not worth while spending more money on a 'good idea'.

Obviously, anyone with a sense of responsibility who sets out to persuade a company or bank, or some other source of funds, that he knows how to develop a new product or process does so convinced that the chances of technical and commercial success are better than even. He would also be justified in believing that the practical value of a new scientific discovery would be sooner, and more critically appreciated in the environment of a research laboratory than it would be in a board-room, where company, and particularly finance directors, would need to be persuaded that it is sensible to provide the resources that would be called for if a project were to go ahead. As the custodians of shareholders' money, it is their responsibility to be wary of the financial risks they take when they back a technological innovation. Ministers have also to be extremely cautious about risking the public's money on major technological projects—for example on a new aero-engine.

But if risks are not taken, there will never be any successes. And if there is to be profitable industrial change, entrepreneurial enthusiasm and balanced technical judgment should never be allowed to become the victim of automatic short-sighted financial scepticism. There is much justification for the criticism that in recent years British industry has been far too cautious in this respect, and far slower in picking up and exploiting new ideas than have the USA, Japan, and Western Germany.

If the big prizes in a world of rapid technological change are to be won, it is necessary to get in at the start—and to remain among the leaders. There is little national benefit in doing the spadework and allowing others to reap the reward. 'Radically new technology has always to *create* its own markets' as has been said by one academic industrialist who has helped in the administration of the Wolfson Scheme, 'and this implies a need for patience, vision and long-term support during the period which people always need to adapt to new opportunities'.[13]

No one need be surprised that because they are publicly accountable, governmental institutions such as the Research Councils and the DTI are necessarily driven to more formal procedures in their negotiations with those university/industry groups which they choose to support than are called for by the Wolfson Foundation. But as the recent ACARD/ABRC Report[12] warns, 'a blinkered approach from the guardians of the public purse will in the long run harm the UK's economic interests far more than any losses from the risks inherent in adopting new technology'. From this point of view the Wolfson Scheme remains unique in so far as it *invites* research workers in universities to bid competitively for funds which the Foundation is prepared to make available, on the condition that, given that their work is successful, they have the promise of subsequent industrial support.

By helping at an early stage, the Scheme thus makes it possible for many to surmount the initial, and therefore the most crucial obstacle in the process of innovation. No doubt some whose ideas have been supported under the Scheme might have found a different way to get started. Today, now that the SERC, the UGC, and DTI are also in the field to encourage direct university/industry collaboration, they might find that the chances of getting started are brighter than they used to be. They might even be helped by the presence of the many science and industry 'parks' that have been set up in recent years in the vicinity of universities, where new businesses or subsidiaries of existing companies are invited 'to set up shop' on favourable terms and, if needed, with the university's promise of technical help or co-operation when called for.

It is plain that there is no shortage of sensible ideas of potential industrial value in British universities. Were it possible to open the scheme to the polytechnics, the choice of projects to support, whether for new products or new processes, could no doubt have been far wider than it has been. Polytechnics clearly have a part to play. The Wolfson Foundation does not have the resources needed to support more than about one in five or ten of the projects now put to it by university groups. Since the country no longer suffers from a shortage of technically trained manpower, it is very much in the national interest that public as well as private money should continue to be made available to help develop direct contacts between universities and industry of the kind that the Foundation had fostered, but in the clear understanding that there can be no guarantee of success whichever way and however extensively links are promoted in the launching of a university/industry project.

In theory, risks might be reduced were support provided only to university scientists and engineers who had proved that the technical case which they were promoting was watertight—if that were ever possible. The risk of failure in setting up a university/industry unit would also be less if support were provided only to those groups which had decided from the outset to provide a straightforward industrial consultancy service, or which had evolved mainly in that direction. But even here there are often difficulties due to over-cautiousness. After the expiry of the initial risk phase which the Foundation has covered, several projects have received help from their parent universities or from SERC—and indeed from firms—when all these other sources were reluctant to provide funds at the start. Caution has, of course, to rule where resources are lacking. But even so, one should never forget that it is the provision of financial help at the start or at an early stage that is all-important. Once started, the units or institutes or companies that have been set up under the Scheme have clearly derived great benefit from the fact that while the members of the units strive to bring a technological idea to fruition they can be in daily touch with university colleagues who are informed about the latest advances that may be relevant to the work they are doing.

Regardless of how projects are started, the main problem in the end is to find the venture capital for projects after they have been independently and successfully pursued to the point where 'real money' is needed to build a prototype plant or to launch what in normal circumstances would be a marketable product—such, for example, as the Manchester motorcycle (p. 23). Judging by results, both the Japanese and Americans have a different and more effective approach to the deployment of risk capital than we have. I suspect, too, that our big manufacturing companies are not as a rule likely to be of much help to small independent university/industry groups. They have their own investment priorities. One is therefore immediately driven to think of the several specialized venture-capital commercial institutions that have been set up in recent years. And the question that needs to be asked is whether they are fully aware of the many university/industrial projects which are now in the pipeline, or which have already established themselves in their own right. Obviously all parties with a possible interest in the success of university/industry groups would undoubtedly benefit if information about the possibilities were more widespread. This small book, which shows the extent and diversity of the Wolfson projects, could help here.

It is to be hoped, too, that the new venture-capital companies are aware of the fact that, in reaching their decisions, they would be better helped if they consulted those who have themselves directed university/industry project teams rather than the many academic pundits who have merely looked in over the fence. British habits in this field need to be changed, and changed quickly. The Government has recently rescinded the monopoly rights of the NRDC and BTG to innovations which they have helped foster. This should help open the field for investment, but it is only one step in the right direction. There are scores of well thought-out innovative ideas waiting to be picked up, even if, once picked up, they then enter the shadows cast by commercial secrecy.

Relatively few of the proposals that have been put to the Foundation were for the support of 'high technology' projects, if by that term is meant subjects such as photo-electronics, semiconductor technology, and genetic engineering. One possible reason for their absence is that government agencies, and some of the largest firms, are already seized of the commercial and, in some cases, military potential of work in these fields. The support of a charitable foundation would be superfluous in such cases. But the diversity of the subjects that we have supported shows that far more than 'high-tech' can be profitable, both in terms of financial rewards and of jobs.

The decline in recent years in the UK's share of the world market for manufactured goods is, of course, due not only to our own inadequacies, but also to the increasingly skilled performance of our competitors. In 1968, the year the Wolfson Scheme was launched, a Government Committee of which I was Chairman published an exhortatory document entitled *Technological Innovation in Britain*.[14] While referring at great length to competition from the United States, Japan and Western Germany were barely mentioned. Today it is Japan which is dominant in the 'high-tech' field. A recent report of the American National Academy of Sciences even goes so far as to talk of the demise (*sic*) of the American consumer electronics industry due to Japanese 'targeting', and of Japan's present bid to dominate the semiconductor market.[15] Economic power in today's technological world can shift very quickly.

Great technological innovations, for example the internal combustion or jet engine, radio, integrated circuits, and the Polaroid camera, are almost as rare as the great ideas which transform science itself—such as the special theory of relativity, quantum mechanics, solid-state physics, and the chemical basis of genetics. Some unique and major technological innovation could crop up in a British university which in the course of time would change our fortunes and way of life, in the same way as did the steam and internal-combustion engines. If such a thing were ever to happen, it is, however, unlikely that its effects would pervade society, and so cure unemployment, any faster than did the invention of steam power. In the meantime, and in order to help smooth the way to Britain's economic recovery, it is simple prudence to encourage the more modest innovative ideas that abound in our universities and technical colleges.

We have practical brain-power to sell. We do not know whether there is any 'best' way of organizing the units and institutes that have been set up under the Wolfson Scheme, any more than their history makes it possible to propose a better way to select the best projects for support than the informed judgment of knowledgeable assessors. As in most fields of endeavour, the success of a university/industrial unit depends in large measure on the quality of those by whom it is launched.

Finally, it should not be forgotten that there are still vestiges in our country of the belief that a career in industy does not merit as much prestige socially as one in the professions, or in what is loosely called 'the City'. This is not a belief that we can afford any longer. If we are to maintain our own standard of living, and help raise that of even poorer countries, we shall have to accept that there is little if any difference in prestige between the qualities that are called for in the search for knowledge for its own sake, and for knowledge that can be put to use. The differences are essentially terminological. They relate mainly to the time-scale in which new ideas, great or small, affect our lives. The basic laws of genetics, and the analytical and chemical procedures of molecular biology, became the foundation stones of a vast biotechnological industry. The discovery of nuclear fission and fusion, Einstein's simple equation $E = mc^2$, blazed the path to nuclear power. The simple microscope and the telescope opened up a world of enquiry which was then illuminated by electron microscopes, by nuclear magnetic resonance body-scanners, by large radar dishes and by space satellites. The 'in' phrase today is 'technology transfer'. The Wolfson Technological Project Scheme has operated to help the process of transfer, in the understanding that transfer takes time, that mankind has always adapted to new knowledge—and that knowledge means power.

APPENDIX: WOLFSON FOUNDATION GRANTS FOR TECHNOLOGICAL PROJECTS

1968

BIRMINGHAM UNIVERSITY

Development of a multi-station petro-forge transfer machine £66 700

CAMBRIDGE UNIVERSITY (A)

Application of modern control theory to models of the national economy £20 000

CAMBRIDGE UNIVERSITY (B)

Research on electron optical instruments £50 000

EDINBURGH UNIVERSITY

Establishment and operation of a micro-electronics liaison unit £130 700

IMPERIAL COLLEGE, LONDON (A)

Investigation of the chemistry of silicates £47 000

IMPERIAL COLLEGE, LONDON (B)

Compilation of a geochemical atlas of part of the British Isles £65 000

LIVERPOOL UNIVERSITY (A)

Research and development to assist manufacturers of vacuum equipment £70 000

LIVERPOOL UNIVERSITY (B)

Unit in engineering design and innovation £15 000

MANCHESTER UNIVERSITY

Studies of the stability of earth works by scale models £7 000

NOTTINGHAM UNIVERSITY

Institute for the study of inter-facial phenomena including research on surfaces between solids, liquids, and gases £255 000

SOUTHAMPTON UNIVERSITY (A)

Establishment of an electronics industrial liaison unit £24 400

SOUTHAMPTON UNIVERSITY (B)

To set up a new centre for industrial noise and vibration control £30 000

SURREY UNIVERSITY

Centre for research and development in bioanalytical instrumentation £132 000

UNIVERSITY OF WALES INSTITUTE OF SCIENCE AND TECHNOLOGY (UWIST)

Centre for the technology of soft magnetic materials and their application £132 000

UNIVERSITY OF WALES, UNIVERSITY COLLEGE, CARDIFF

Extension of research into the mineral reserves other than coal, of south and mid-Wales £10 800

1970

ABERDEEN UNIVERSITY

Research project to develop kenaf as a substitute for jute £17 500

ASTON UNIVERSITY

The evaluation of metals, alloys, and other materials in terms of their ultimate cost per unit of property, to establish criteria involving the economic and technical factors for the reduction of materials costs £17 630

BELFAST, QUEEN'S UNIVERSITY

The establishment of an Opto-Electronics Centre which would make available to Industry the research capabilities of the laser and opto-electronics research group £75 000

BIRMINGHAM UNIVERSITY

A broad investigation into improvements in processes for the recovery of non-ferrous metals from secondary sources such as scrap and waste £128 200

NEWCASTLE UNIVERSITY (A)

Establishment of a research and development group for the production and characterization of new metallic and non-metallic materials £155 850

NEWCASTLE UNIVERSITY (B)

Extension of work on research into rock breakage under the action of powered mechanical tools, with the aim of designing better cutting systems for mechanical tunnel excavation £76 800

ST. ANDREWS UNIVERSITY

Founding of an institute of luminescence. Work to be concentrated initially on electroluminescence £75 000

SOUTHAMPTON UNIVERSITY (A)

Expansion of engineering materials advisory service £50 000

SOUTHAMPTON UNIVERSITY (B)

Establishment of an Applied Electronics Advisory Unit £24 800

SOUTHAMPTON UNIVERSITY (C)

Expansion of the Marine Craft Advisory Unit £16 500

UNIVERSITY OF WALES, UNIVERSITY COLLEGE, CARDIFF (A)

Establishment of a Laboratory of Biology in Industry which will study the effects of living organisms in processes such as steel manufacture and the use of microbes to industry £40 000

UNIVERSITY OF WALES, UNIVERSITY COLLEGE, CARDIFF (B)

Continuation of research into mineral reserves, other than coal, in Wales (made possible by grant from Foundation in 1968). Also to provide for both specialized training and technical service to mining and smelter firms £64 450

1972

ASTON UNIVERSITY

The establishment of a Heat Treatment Advisory Centre £75 000 over three years

BATH UNIVERSITY

Development of a laboratory for engine transmission system investigations £92 000 over three years

BELFAST, QUEEN'S UNIVERSITY

To form a composite construction group in the Faculty of Applied Science to exploit the extra scope in metal construction arising from the advent of resins, strong fibres, and foamed solids 54 000 over five years

CAMBRIDGE UNIVERSITY

Study of phase transitions
in dielectric materials £39 400
(electrical insulators) over five years

CITY UNIVERSITY

Development of a Unit of
Electrochemical £36 050
Technology over three years

DUNDEE UNIVERSITY

Research into the improvements to road decks £99 000
of bridges over six years

EDINBURGH UNIVERSITY

To establish a unit for high-
speed liquid chromatography
(form of chemical £92 500
analysis) over five years

ESSEX UNIVERSITY

Support for the Electronics
Centre in linking it with £33 000
local industry over three years

HERIOT-WATT UNIVERSITY

Establishment of an Institute
of Offshore £177 790
Engineering over four years

IMPERIAL COLLEGE, LONDON

Research into methods of
speeding up high-tempera-
ture metallurgical £72 609
processes over five years

LEEDS UNIVERSITY

Establishment of a research
unit for the study of the formation of technical organic
powders and their physical £84 278
properties over five years

LOUGHBOROUGH UNIVERSITY

Research to develop techniques
to reduce the cost of producing construction
information—Project £17 500
'LUCID' over two years

NEWCASTLE UNIVERSITY

To establish a Chair of
Energetics which would
take a comprehensive
view over the whole field £37 500
of energetics over five years

NOTTINGHAM UNIVERSITY

To establish a research and
development group to
specialize in the further
development and under-
standing of mechanical £44 560
assembly techniques over three years

SUSSEX UNIVERSITY

Research to develop a
magnetically suspended
vehicle for urban £127 655
transportation over three years

WARWICK UNIVERSITY

To design and construct
a magnetically levitated
test vehicle and track
using superconducting £147 750
magnets over five years

1974

ASTON UNIVERSITY

Impact of motorways and
other new principal road
schemes on agriculture £28 720

CRANFIELD INSTITUTE OF TECHNOLOGY

The combustion of residual
fuel oils in gas turbine
systems £155 000

HERIOT-WATT UNIVERSITY

Energy research: utili-
zation, directly and in-
directly, of solar energy £20 000

IMPERIAL COLLEGE, LONDON (A)

The production of smectites from unconventional UK sources   £60 000

IMPERIAL COLLEGE, LONDON (B)

The extraction of metals from low-grade sources by coupled molten halide solvation and electrochemical reduction   £106 000

INTERNATIONAL SOLAR ENERGY SOCIETY

Preparation of a report on the potential of solar energy utilization in the UK   £5 000

LIVERPOOL UNIVERSITY (A)

Scallop cultivation: its feasibility and economic viability   £6 547

LIVERPOOL UNIVERSITY (B)

Development of an efficient process for the reclamation and recycling of thermoplastics from contaminated waste   £61 840

QUEEN MARY COLLEGE, LONDON

Survey of the processes available for the recycle of plastics and paper   £93 000

MANCHESTER UNIVERSITY

Fast-flow biomass production from dilute food-processing wastes   £68 300

NATIONAL COLLEGE OF AGRICULTURAL ENGINEERING

Straw utilization: a techno-economic study of straw collection, storage, and transportation from farm to procession factory   £13 700

OXFORD UNIVERSITY

Bacterial action in re-cycling metals   £17 400

READING UNIVERSITY (A)

Leaf protein: production and utilization   £120 000

READING UNIVERSITY (B)

Production and utilization of oil and protein from seed crops   £128 568

STRATHCLYDE UNIVERSITY

Methane energy systems using fermentation of organic wastes   £27 578

SURREY UNIVERSITY

Improvements in papers and inks   £39 000

UNIVERSITY OF WALES, UNIVERSITY COLLEGE, ABERYSTWYTH

Extending the growing season and increasing the over-wintering ability of legumes and grasses in upland areas   £14 300

YORK UNIVERSITY

Optimization of sludge-digestion processes   £44 050

1976

ABERDEEN UNIVERSITY

Development of an electrochemical discharge process for deburring metal components   £71 500 over five years

ASTON UNIVERSITY

Wolfson Industrial Ultrasonic Material Forming Units   £100 000 over three years

BATH UNIVERSITY

Land mobile radio systems £102 000
using SSB over three years

BELFAST, QUEEN'S UNIVERSITY

Human and industrial £52 500
signal processing over five years

INSTITUTION OF CHEMICAL ENGINEERS

Chemical engineering in £83 000
agriculture over three years

LIVERPOOL UNIVERSITY

Plasma processing of £120 000
tools over three years

IMPERIAL COLLEGE, LONDON

New pathways of indus- £46 400
trial fermentation over three years

QUEEN MARY COLLEGE, LONDON

Sea bed anchorage £238 000
over three years

MANCHESTER UNIVERSITY (A)

Electronic monitoring £26 500
textile processes over two years

MANCHESTER UNIVERSITY (B)

Image measurement systems
for industrial and bio- £156 000
medical applications over three years

NEWCASTLE UNIVERSITY

Thin-film device develop-
ments for electronic pre- £155 000
cision applications over four years

NOTTINGHAM UNIVERSITY

Research on NMR body £118 000
scanner over three years

OXFORD UNIVERSITY (A)

Development of a Neo-
natal Bio-Engineering £107 600
Unit for five years

OXFORD UNIVERSITY (B)

Development of high-
performance artificial £229 000
kidney over four years

ST. BARTHOLOMEW'S HOSPITAL, LONDON

Development of medical £184 000
electronic equipment over five years

SALFORD UNIVERSITY (A)

Establishment of an Ion £96 000
Plating Unit over two years

SALFORD UNIVERSITY (B)

Development of energy £75 000
storage system over three years

SOUTHAMPTON UNIVERSITY (A)

Research on the use of
microprocessors in the
control of manufac- £25 500
turing processes over three years

SOUTHAMPTON UNIVERSITY (B)

Research on thin polymer
films for videodisc £48 000
applications over three years

SOUTHAMPTON UNIVERSITY (C)

Development of a one
million r.p.m. motor-
driven spindle for the £82 000
textile industry over three years

SOUTHAMPTON UNIVERSITY (D)

Research of chemicals for
use in insect pest con- £136 000
trol over four years

STIRLING UNIVERSITY

Establishment of a Unit of £68 400
Seed Technology over three years

STRATHCLYDE UNIVERSITY

Development of Prototype
SF6 puffer circuit £109 500
breaker over three years

1978

UNIVERSITY OF WALES, UNIVERSITY
COLLEGE, BANGOR

Computer-aided design of
microwave integrated
circuits (Radar and   £55 950
communications)   over three years

BATH UNIVERSITY (A)

Investigation into a con-
tinuous process for
edible bone protein.
(Animal food/meat
extracts for human   £49 400
food)   over two years

BATH UNIVERSITY (B)

Development of a range of
surface stressed flexible
structures. (Textiles in   £159 900
building construction)   over three years

BATH UNIVERSITY (C)

A new diesel-based power-
plant system for heavy   £50 000
commercial vehicles   over two years

BELFAST, QUEEN'S UNIVERSITY

The utilization of mollusc   £49 450
resources (cattle food) over four years

BIRMINGHAM UNIVERSITY (A)

Further developments in
composting (sewage to
horticultural/mushroom   £163 700
compost)   over four years

BIRMINGHAM UNIVERSITY (B)

Setting up of an Industrial
Unit for the bulk form-
ing of materials. (Cross-
fertilization, between
industries, of metal-   £120 000
shaping techniques)   over two years

BIRMINGHAM UNIVERSITY (C)

To establish an underwater

acoustics instrumenta-
tion facility. (Underwater   £95 000
structure inspection)   over five years

DUNDEE UNIVERSITY

Device applications of
amorphous semicon-
ductors. (Solar energy
conversion and indus-   £35 000
trial displays)   over one year

IMPERIAL COLLEGE, LONDON (A)

Proposal to establish a
Solid State Ionics Unit
and to develop specific
solid state ionic devices.
(Control of chemical
plant/solid-state   £125 000
batteries)   over two years

IMPERIAL COLLEGE, LONDON (B)

New electron-optical
streak tubes for pico-
second chronoscopy and
the picosecond oscillo-
scope. (Optical data
processing and com-   £75 000
munications)   over three years

MANCHESTER UNIVERSITY

Establishment of a Motor-   £183 660
cycle Research Unit   over four years

NEWCASTLE UNIVERSITY

Gasification of petroleum
coke and factors influ-
encing mechanical and
chemical properties of
resultant graphites in
industrial environments.
(Improvements in graph-
ites for arc-electrodes   £17 700
and nuclear reactors)   over three years

SHEFFIELD UNIVERSITY (A)

The investigation of elec-
trocoating mechanisms
and the computer-aided
design of jigs. (More

uniform chrome plating) £40 000 over two years

SHEFFIELD UNIVERSITY (B)

Plant cell cultures and their biotechnological application. (Biological routes to pharmaceuticals) £120 000 over three years

SOUTHAMPTON UNIVERSITY

Research and development centre to provide a comprehensive service to industry in the fields of electrochemistry and electrochemical engineering £162 660 over four years

STIRLING UNIVERSITY

Further support for the Wolfson Unit of Seed Technology for work on seed production £37 800 over two years

STRATHCLYDE UNIVERSITY

Development of an improved type of shaded-pole, single-phase induction motor. (Cheaper designs of small motors) £106 170 over three years

UNIVERSITY OF ULSTER, COLERAINE

Development of a biological system for the treatment and utilization of whiskey distillery waste £140 580 over three years

UNIVERSITY OF MANCHESTER INSTITUTE OF SCIENCE AND TECHNOLOGY (UMIST)

Textile design conversion system. (Design taken via computer to setting-up knitting machines/looms) £44 800 over two years

YORK UNIVERSITY

Infrared analysers for quality and process control in the food industry £100 000 over three years

1980

ABERDEEN UNIVERSITY (A)

Ultrasonic probes of high resolution (for non-destructive testing and application in medicine) £50 000 over three years

ABERDEEN UNIVERSITY (B)

Improvement of birch planting stock by selection and breeding £61 000 over five years

BATH UNIVERSITY (A)

Custom-design of LSI microelectronic circuit £127 000 over three years

BATH UNIVERSITY (B)

Development of microbial N-dealkylation processes in the manufacture of pharmaceutical raw materials £100 000 over three years

BIRMINGHAM UNIVERSITY

Thermal imaging of integrated circuits £110 000 over one year

CAMBRIDGE UNIVERSITY

Treatment of fresh fruit and vegetables £156 000 over four years

UNIVERSITY OF WALES, CARDIFF

Project managers for University College Cardiff Industry Centre £54 000 over three years

CRANFIELD INSTITUTE OF TECHNOLOGY

Development of ware-
house order-assembly   £52 000
machine   over two years

ESSEX UNIVERSITY (A)

Contracts with local   £39 000
industry   over three years

ESSEX UNIVERSITY (B)

Unit for solving indus-   £100 000
trial noise problems   over three years

ESSEX UNIVERSITY (C)

Proposal to introduce
small computer systems
to local commercial and   £60 000
trading concerns   over five years

IMPERIAL COLLEGE, LONDON

A support unit, for the
industrial application of   £170 000
microprocessors   over five years

KENT UNIVERSITY

Genetic manipulation of
the gene for talin (artifi-   £90 000
cial sweetener)   over three years

LIVERPOOL UNIVERSITY

Creation of a research and
development advisory   £69 000
service   over five years

MANCHESTER UNIVERSITY

Establishment of an Indus-
trial Advisory Unit in
Maintenance-related
Management and   £190 000
Engineering   over four years

UNIVERSITY OF MANCHESTER INSTITUTE OF
SCIENCE AND TECHNOLOGY (UMIST)

Development of new
materials and prototype
machinery offering
energy conservation in

polymer production and   £113 000
fabrication   over three years

NOTTINGHAM UNIVERSITY

The development of a
photo-conductor for use
in 'intelligent' copying   £136 000
machines   over three years

SALFORD UNIVERSITY

Improvement in picking
systems in warehousing   £105 000
and distribution   over three years

STRATHCLYDE UNIVERSITY

Soy sauce—production
and marketing studies of
an improved fermen-   £88 000
tation process   over three years

UNIVERSITY OF ULSTER, COLERAINE

Establishment of an Industrial Consultancy Service
at the New University   £70 000
of Ulster   over three years

YORK UNIVERSITY

York Electronics   £60 000
Centre   over three years

1981

BATH UNIVERSITY

The development of techniques for vertical position
sensing and control of   £130 000
coal-cutting machines   over three years

BELFAST, QUEEN'S UNIVERSITY (A)

A speech-accessed data   £80 000
base   over three years

BELFAST, QUEEN'S UNIVERSITY (B)

Small-scale commercial
applications of wave   £80 000
power conversion   over three years

CAMBRIDGE UNIVERSITY

Prediction of corrosion in
complex service con-   £60 000
ditions   over three years

CITY UNIVERSITY

Research and develop-   £75 000
ment in gas detection   over two years

CRANFIELD INSTITUTE OF TECHNOLOGY

Automotive Impact Centre   £120 000
                             over two years

EDINBURGH UNIVERSITY

Novel communications
system offering improved
safety for deep-sea   £135 000
divers   over three years

IMPERIAL COLLEGE, LONDON

The development of chemi-
cally modified elec-   £100 000
trodes   over three years

LOUGHBOROUGH UNIVERSITY

Automation of certain
sock manufacturing pro-   £90 000
cesses   over three years

MANCHESTER UNIVERSITY

The establishment of a
Unit for the design and
prototypical development
of components equip-   £60 000
ment and furnishings   over three years

NEWCASTLE UNIVERSITY (A)

The production and characterization of hard, high-
strength materials for
cutting tools, abrasives,
and other engineering   £100 000
applications   over three years

NEWCASTLE UNIVERSITY (B)

On-line monitors for plant
process control and pro-   £125 000
tection   over three years

OXFORD UNIVERSITY (A)

Membrane lungs for open-
heart surgery and the
treatment of respiratory   £75 000
distress   over three years

OXFORD UNIVERSITY (B)

A new consolidation test   £30 000
                             over two years

QUEEN MARY COLLEGE, LONDON

Suppression of smoke and
toxic gases from burning   £140 000
polymer foams   over three years

SALFORD UNIVERSITY

New Products Group   £140 000
                             over two years

SHEFFIELD UNIVERSITY

Engineering feasibility—   £50 000
Winslow effect fluids   over two years

SOUTHAMPTON UNIVERSITY (A)

Industrial Advisory Unit   £100 000
in cryogenics   over three years

SOUTHAMPTON UNIVERSITY (B)

Auditory Communication
and Hearing Conser-   £120 000
vation Unit   over three years

SUSSEX UNIVERSITY

Electromagnetic bearings
for machine tool   £80 000
applications   over two years

UNIVERSITY COLLEGE, LONDON (A)

Instrumentation for the
detection and prevention   £50 000
of brain damage   over three years

UNIVERSITY COLLEGE, LONDON (B)

Wolfson Unit for Micro-   £70 000
NDE   over four years

UNIVERSITY COLLEGE, SWANSEA

A high-voltage integrated
circuit process and its £200 000
application over three years

WARWICK UNIVERSITY

The design of high-
performance ceramic £80 000
alloys over three years

1982

BIRMINGHAM UNIVERSITY

Surface Technology £100 000
Support Unit over three years

CAMBRIDGE UNIVERSITY

Fluid flow and mixing in
industrial flow £170 000
processes over three years

HERIOT-WATT UNIVERSITY

Radio-frequency excited
waveguide laser £90 000
system over three years

HULL UNIVERSITY

Flexible automatic £50 000
handling systems over two years

IMPERIAL COLLEGE, LONDON

Total synthesis of novel
anthelmintic £200 000
substances over three years

IMPERIAL COLLEGE, LONDON

Soft X-ray instrumen- £200 000
tation and optics over three years

LEEDS UNIVERSITY

Computer-aided designs £200 000
for organic syntheses over three years

LIVERPOOL UNIVERSITY

Development of electronic
systems for tungsten £100 000
inert gas welding over three years

MANCHESTER UNIVERSITY

Development of hardware
and software for indus-
trial inspection £160 000
and robot vision over three years

OXFORD UNIVERSITY

High power copper £180 000
laser development over three years

OXFORD UNIVERSITY

Cheaper substitute
material based on £75 000
cement over three years

READING UNIVERSITY

Optical design and con- £80 000
sultancy service over three years

SALFORD UNIVERSITY

Dynamic recoil mixing £140 000
unit over two years

SOUTHAMPTON UNIVERSITY

Mosquito control by
insoluble monolayers £120 000
and bilayers over three years

SURREY UNIVERSITY

Continuous process for
purification of £120 000
antibodies over three years

UNIVERSITY COLLEGE, LONDON

Optical fibre instrumen- £100 000
tation systems over four years

UNIVERSITY COLLEGE, LONDON

Three-dimensional mea-
surement systems for £40 000
industry over two years

1983/4

BELFAST, QUEEN'S UNIVERSITY

Development of vibration £35 000
arthrograph over three years

BIRMINGHAM UNIVERSITY

Component production by   £123 000
  preform process   over three years

BRUNEL UNIVERSITY

Injection moulding of   £97 000
  engineering ceramics   over two years

CAMBRIDGE UNIVERSITY

Novel anti-arthritic   £75 000
  drugs   over three years

ESSEX UNIVERSITY

Design and control of   £100 000
  dynamic structures   over three years

GLASGOW UNIVERSITY

Development of thermo-   £103 000
  tolerant streptomyces over three years

LEEDS UNIVERSITY

Combustion monitoring   £125 000
  and control   over three years

LIVERPOOL UNIVERSITY

Development of adhesive   £142 000
  dental materials   over four years

IMPERIAL COLLEGE, LONDON

Zeolite catalysts in   £185 000
  detergents   over three years

IMPERIAL COLLEGE, LONDON

Constructional ceramics
  from industrial   £175 000
  by-products   over two years

QUEEN MARY COLLEGE, LONDON

New anti-adhesion   £100 000
  coatings   over three years

MANCHESTER UNIVERSITY

Packaging for sterile   £145 000
  items   over three years

MANCHESTER UNIVERSITY

Liquid crystal thermo-   £128 000
  metry   over three years

MANCHESTER UNIVERSITY, INSTITUTE
OF TECHNOLOGY

Development of on-line   £48 000
  monitors   over two years

NOTTINGHAM UNIVERSITY

Active photoconductor   £173 000
  systems   over three years

SOUTHAMPTON UNIVERSITY

Electrostatic explosion   £75 000
  hazards   over three years

UNIVERSITY OF WALES, BANGOR

Development of magnetic   £60 000
  ink   over three years

WELSH NATIONAL SCHOOL OF MEDICINE

Elimination of ethylene   £50 000
  oxide   over three years

YORK UNIVERSITY

Captor drum waste water   £72 000
  treatment   over two years

REFERENCES

1. Committee on Higher Education *Higher Education* (Cmnd. 2154). HMSO, London (1963).
2. University Grants Committee *University development 1962-1967* (Cmnd. 3820). HMSO, London (1968).
3. Wood, Sir Henry P. Presidential Address at Dundee Meeting of the British Association for the Advancement of Science (1968).
4. Science Research Council *Report of the Council for the year 1965-66* (HC 203). HMSO, London (1966).
5. Science Research Council *Report of the Council for the year 1970-71* (HC 517). HMSO, London (1971).
6. Select Committee on Science and Technology Third Report. *University-Industrial Relations.* (HC 680). HMSO, London (1976).
7. Zuckerman, S. Proposals for a review of policy in the fields of Higher Education. Wolfson internal memorandum (1967).
8. Select Committee on Science and Technology (1982) *Science and Government,* II—Evidence. HMSO, London (1982).
9. Harvey-Jones, J. *Sunday Times* (18 September 1983).
10. *Annual Review of Government Funded R & D 1983.* HMSO, London (1984).
11. Advisory Council for Applied Research and Development. *Improving research links between higher education and industry.* HMSO, London (1983).
12. Advisory Board for the Research Councils and Advisory Council for Applied Research Development. *Joint Reports by the Chairman,* 1 (Cmnd. 8957). HMSO, London (1983).
13. Smith, D. *The Times* (17 September 1983).
14. Central Advisory Council for Science and Technology. *Technological innovation in Britain.* HMSO, London (1968).
15. American National Academy of Sciences. *International competition in advanced technology: decisions for America.* National Academy Press, Washington D.C. (1983).